RAPTOR PREY
REMAINS

RAPTOR PREY REMAINS

A Guide to Identifying What's Been Eaten by a Bird of Prey

ED DREWITT

PELAGIC PUBLISHING

Published by Pelagic Publishing
PO Box 874
Exeter
EX3 9BR
UK

www.pelagicpublishing.com

Raptor Prey Remains: A Guide to Identifying What's Been Eaten by a Bird of Prey

ISBN 978-1-78427-207-4   *Paperback*
ISBN 978-1-78427-208-1   *ePub*
ISBN 978-1-78427-209-8   *PDF*

A CIP record for this book is available from the British Library

Cover photographs:
Sparrowhawk with starling, Terry Stevenson
Barn owl feathers, Sophie Bagshaw
Blackbird head, Saimon Clark
Teal feathers, Ed Drewitt

Printed and bound in India by Replika Press Pvt. Ltd.

To my wonderful wife Liz, who also loves wildlife, and to Freddie, who already loves being outdoors and spotting animals, plants and signs of nature.

Contents

Introduction

....

Raptors, also known as birds of prey, take a surprisingly wide range of prey species. Discovering what they have been eating provides important insights into the behaviour and ecology of both predator and prey, although identifying the prey can be a challenge. This book is a compilation of the most commonly encountered species taken by raptors in our towns, cities and countryside. It deals with the discarded or stored (cached) remains of animals and includes photos of over 100 prey species, representing how their remains are most often found at raptor nests, roosts, plucking posts and other opportunistic spots. Whether you have just found a pile of feathers on your lawn, or you have collected prey remains from a plucking area, this book aims to help you take the first steps towards working out what your discovery is. It enables you to piece together the clues, be a nature detective, and come up with a positive identification that can be compared with others in a book, museum or website.

I have endeavoured to show what you are most likely to find from a particular bird or mammal when it has been taken by a raptor. The bird species are shown in taxonomic order according to the current (2019) British Ornithologists' Union list (bou.org.uk). I have not shown any scales with the photos, because of the complexity of sourcing images and also because I want this to be a visual guide. There are many other resources that deal with measurements in greater detail, and these can be useful when a feather or skull needs further scrutiny to be identified to a species. This book deals mainly with birds, as many books already adequately cover the identification of other animals, particularly small mammals. I have kept 'other animals' mainly visual, providing images of how common prey such as rabbits and grey squirrels are found.

Finding and identifying raptor prey remains

WHERE TO BEGIN

The remains of animals that a raptor has taken are found in all manner of sites, from nest boxes to open fields. Some are found at favourite plucking places or nests, while others may have been plucked and eaten opportunistically in an open space or within a woodland. Where a single animal, for example a duck or rodent, has been taken, photographing or collecting as much of the evidence as possible will help in identifying the species. Sometimes a particular feather, a claw or an arrangement of teeth can really help clinch the identification.

WHO'S BEEN AT WORK? A MAMMAL OR BIRD PREDATOR?

When you find a scene of feathers, fur and other body parts, a closer examination can reveal whether a bird or mammal predator has been at work. Often prey remains are found at known roost or nest sites. However, some may be found away from places which raptors use regularly.

Signs of a raptor kill

Feathers plucked by raptors will remain whole, although they may be slightly bent or show a crease or kink in the rachis; some of the feather vane may also be torn and the beginning of the rachis – where it fits into the bird's skin – may show a darkening tip where a recent bloody supply has dried. Many feathers may show little sign of damage. The fur of a mammal, such as a rabbit, will be scattered in small, light fluffy piles, and the skull will often be picked clean. The cranium of the skull may be opened up in one area, allowing the raptor to eat the brain, which is rich in fatty acids.

Signs of a mammal kill

If a mammal has been the predator or scavenger, there are often telltale signs of dried saliva that binds small feathers or fur together, and resembles the trails left by slugs and snail. The wing and tail feathers are snipped off, as if someone has used a pair of scissors. Body parts, such as legs, heads and the body itself may be well chewed and mashed up. Scats (poo) may be found nearby, sometimes containing some of the same feathers or fur, as well as scent left by a mammal such as a fox.

It isn't always possible to confirm whether the prey has been caught by a raptor first and then scavenged by a mammal, or whether it was caught by a mammal in the first place. In urban locations, prey dropped by peregrines may be secondarily scavenged by cats, foxes or badgers. If a headless carcass is found on the ground, either a bird predator, such as a sparrowhawk, or a mammal, such as a fox, might be responsible.

Damage to a cuckoo feather by the beak of a bird of prey – the shaft is kinked and flattened

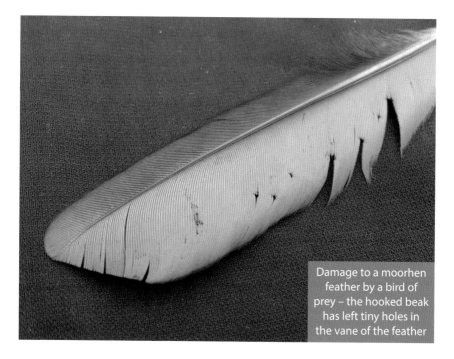

Damage to a moorhen feather by a bird of prey – the hooked beak has left tiny holes in the vane of the feather

Pile of plucked collared dove feathers, typical of a bird of prey such as a sparrowhawk VM

Pile of feathers of a predated feral pigeon BS

Feather remains of a blackbird from a sparrowhawk kill

VISITING REGULAR PREY SITES

Sites such as roosts or breeding locations may provide a regular supply of prey remains. When beginning a diet study at known regular prey sites it is worthwhile collecting everything you find on a first visit; thereafter anything else you find is likely to be new. Just be aware of old feathers and skulls being washed down gutters and collecting over drain covers, and that some prey may be stored for later eating (and further plucking). Visit as regularly as you can, whether that is daily, weekly, monthly or seasonally. Weekly or more often is ideal, as the small feathers of birds or wings of bats are more likely to be found before they are blown or swept away.

Pellets

The pellets of raptors can be useful for studying their diet, particularly for those eating birds, small mammals or other animals whole. This book deals with the discarded or stored (cached) remains of animals found rather than those which have been eaten and their bones, fur or feathers regurgitated in pellets. For identifying bones from pellets, this book is a useful guide:

- Yalden, D.W. *The Analysis of Owl Pellets*, 4th edition. The Mammal Society, Southampton, 2009.

COLLECTING SAFELY

Some remains are likely to be bloody and maggoty; it is advisable to wear disposable gloves to avoid picking up any diseases and for good hygiene practice. Keep hand sanitiser or wipes to hand and wash hands properly with soap and water at the earliest opportunity. Dry feathers and skulls can be stored and labelled in sealable plastic bags. Wings and legs may need drying for several weeks. Even during the winter months heads and corpses can have maggots or other insect larvae in them; this can be ideal if you are wanting to get the skull. Heads can be left outdoors in a covered small plant pot to decompose (and away from where foxes and other animals can find them).

Keep records such as date, location and species of prey items. This helps you to refer back to them, and to provide information to county recorders – and it can also be important evidence of where you collected them from a legal perspective.

Legalities of prey remains

While in the UK it is legal to keep feathers and skulls of some species, it may not be legal to do so in other countries – in which case you may need to dispose of your finds, liaise with a local museum or university, or seek permission or obtain a licence.

In the UK, you need to check whether the animal is subject to Schedules 1 to 10 of the Wildlife and Countryside Act (as amended) (legislation.gov.uk/ukpga/1981/69). If the species is listed, contact the licensing department of Natural England, Scottish Natural Heritage, Natural Resources Wales or the Northern

Ireland Department of Agriculture, Environment and Rural Affairs for advice on whether you are allowed to keep body parts (gov.uk/guidance/wildlife-licences). Take photos and records of where the remains were found, and ideally a photo of them in situ. If there are any suspicions around the find, please contact your local law enforcement for guidance before removing the remains from the scene.

The CITES convention (cites.org) governs worldwide trade in endangered species, and at the time of writing there are 183 member parties, with many countries having adopted the convention into law. Animals (dead, alive or body parts) listed under CITES are strictly regulated. If a species is subject to CITES please contact the relevant country's governing body, which in the UK is DEFRA (gov.uk/guidance/cites-controls-import-and-export-of-protected-species), before buying, collecting or storing.

WORKING OUT HOW MANY INDIVIDUALS OF A SPECIES YOU HAVE

When you are collecting lots of prey remains, the challenge can be to work out what is from a single bird of a species, and what is from more than one individual. A single bird may have thousands of feathers. The primary, secondary and tertial feathers can help with determining whether one or more individuals have been eaten. Counting legs, heads (or parts of heads) and wings will also confirm. When you have a pile of feathers, separate them out into their different species. Then, within each species, look to see if there are any feathers that appear to be from the same part of the wing or tail. For example, the common grey form of the feral pigeon has a distinctive outer primary feather which has a narrow outer web with a white fringe. If you have more than one such feather from the same-side wing, you have more than one pigeon. Their outer tail feathers, if found, often have an almost complete white outer web which is missing in the other tail feathers. If in doubt, record the very minimum number of individuals taken.

If collecting prey from a site daily, you may encounter the feathers or other parts of the same bird for several days to come as they get blown down from above, or as a cached prey item is plucked further. The feathers and parts of individual birds might even be found over a period of several weeks. To avoid double counting, look for the freshness of the remains and whether there are any duplicates of body parts or particular feathers. Older feathers often look a little more ragged, especially if there has been some rain. Feathers that are several months old – especially during the autumn and winter months – will begin to decay a little, developing a dusty white matt colour to the rachis; the webs of the feather will often look weathered and dirty too.

BUILDING UP A REFERENCE COLLECTION AND PROTECTING FROM INSECTS

Alongside keeping feathers and skulls mixed in storage bags or boxes related to sites and dates, a refence collection of species can be developed. For feathers I have seen a range of solutions, for example delicately scalpel-scored card pages forming slits for feathers to fit through. However, sealable bags that can keep pests out are ideal, and they can be stored in boxes.

Feathers and fur attract invertebrates that want to eat them, particularly the larvae of the varied carpet beetle, which look like small hairy maggots. When storing dry samples such as feathers, put them in a sealable bag and leave them in the freezer for a month. Keep an eye on samples over months and years; if any show signs of being eaten, put them back in the freezer. Signs include holes appearing, fragments of feathers detached from the feather, and minute brown particles which are the insects' poo.

WHAT CLUES DO DIFFERENT RAPTORS LEAVE BEHIND?

Raptors eat a wide range of animals and employ many different hunting techniques. Some catch their prey through impressive aerial encounters or brief opportunistic flights where the element of surprise is key. Others feed on carrion, animals that are already dead, while others specialise in certain seasons on slower, more sluggish prey such as earthworms and beetles. Some locations such as open moorland and wetlands attract a range of raptors, and several species could be responsible for a kill. For example, on a wetland reserve, the feather remains of a meadow pipit may be from a sparrowhawk or merlin kill, while the remains of a coot could be from a marsh harrier or a peregrine. Despite this, whatever raptor is involved, there are often telltale signs that give away the identity of the hunter. Below are the general prey-finding characteristics of some raptor species. Those marked with an asterisk (*) are on Schedule 1 of the Wildlife and Countryside Act, which means that in the UK you must have a licence issued by the British Trust for Ornithology or the relevant Country Agency before visiting an active nest or disturbing dependent young.

Golden eagle *Aquila chrysaetus**

The prey remains are most likely to be found at the nest, and range from body parts of foxes to mountain hares, squirrels and remains of birds such as crows. Golden eagles will eat corpses of larger mammals such as deer. These will be eaten in situ, often in the company of buzzards and ravens *Corvus corax*.

Sparrowhawk *Accipiter nisus*

During and after feeding, sparrowhawks leave a pool of feathers and entrails on a garden lawn or park. They often have favourite plucking posts in woodlands too, using particular branches or stumps to pluck prey, and feathers are also often found beneath nest sites. During spring and summer, woodland walks may reveal a pile of feathers from a fledgling bird such as a blue tit, freshly plucked by a sparrowhawk. Birds commonly found plucked by sparrowhawks on lawns and parks include collared doves, woodpigeons, blackbirds and starlings.

Goshawk *Accipiter gentilis**

You are most likely to encounter the prey of a goshawk when visiting a nest. They don't have reliable plucking posts, although in the vicinity of the nest site pellets, feathers, bones and fur may be found. Most nests are kept clean and may show no sign of any prey, while others may be littered with the remains of rabbits, squirrels and birds. Goshawks may sometimes bring prey down in larger gardens, and in mainland Europe, for example in Germany, they are common predators in city parks close to people.

Hen harrier *Circus cyaneus**

Hen harriers are ground-nesters and will bring plucked prey to the nest. They eat mainly birds and small rodents. Piles of feathers found on the ground or stuck in heather in open moorland, beneath posts or fences and along paths may be from hen harrier, although other raptors such as merlin, sparrowhawk and peregrine cannot be discounted.

Buzzard *Buteo buteo*

As with the goshawk, most nests are kept clean of prey although some may be littered with the remains of animals such as rabbits. Parts of birds, especially feathers, are also detected – many may be from road kills that have been scavenged rather than taken alive by the buzzards. Corpses of larger mammals such as deer, sheep and foxes will be eaten in situ and often attract other scavengers such as ravens and perhaps golden eagles in the appropriate locations.

Barn owl *Tyto alba**

The prey of barn owls is most commonly detected through dissecting their pellets, and generally involves small mammals, although birds such as starlings will also feature. Bird remains, such as wings, legs and heads, may also be found in the nest box and beneath or near the nest. A range of species may be detected, including water rail and snipe. Adult barn owls often cache small mammals in the nest.

Tawny owl *Strix aluco*

Like the barn owl and kestrel, the tawny owl feeds primarily on small mammals such as mice and voles, which can be detected by dissecting pellets from nest locations or beneath roost sites. Whole or half-eaten small mammals may be found in nest boxes. Feathers of small to medium-sized birds may also be found in the nest box.

Kestrel *Falco tinnunculus*

Kestrels primarily hunt and eat small mammals, including bats, although birds also feature. Whole or parts of small mammals are found in the nest, while pellets are found beneath entrance holes and favourite perching posts or branches. Feathers and other parts of birds may also be found in the nest.

Merlin *Falco columbarius**

Like hen harriers, which share similar habitat, merlins are ground-nesters and will bring plucked prey to the nest. They eat mainly birds, and piles of feathers found on open moorland – on the ground, amongst heather, beneath fence posts or along paths – may also be from a merlin kill, although other raptors cannot be discounted.

Peregrine *Falco peregrinus**

The food remains of peregrines are frequently found beneath urban locations where they are roosting or breeding. Fresh heads, legs, wings and feathers alongside older skulls and corpses may be found below or on the roofing areas. Scanning around the gutters, drains, paths, car parks and kerbsides within 30 metres of an urban peregrine site is likely to be productive; some feathers may float even further afield. As well as prey remains at or near nest sites in rural locations, prey remains of individual birds or occasionally mammals such as young rabbits may be found alone on exposed grassy places or along the edge of sea cliffs. Prey may be brought down and plucked in situ in open fields and wetlands, and occasionally in large open gardens or parks.

OTHER RESOURCES FOR IDENTIFYING PREY

There are a range of next steps and further resources that can help with examining your remains in more detail. A few of those listed below may be out of print, but still available in libraries or as second-hand copies.

- Brown, R., Ferguson, J., Lawrence, M. and Lees, D. *Tracks and Signs of the Birds of Britain and Europe*, 2nd edition. Christopher Helm, London, 2003. Beautiful illustrations of many species you are likely to encounter, and measurements of primary, secondary and tail feathers.

- Cieślak, M. and Bolesław, D. *Feathers: Identification for Bird Conservation*. Natura Publishing House, Poland, 2006. This book covers European raptors, and birds with similar and striking patterns such as waders.

- The Featherbase website (featherbase.info) details the feathers of European species and lays them out in wing or tail order. This is especially useful once you have an idea of what your remains may have come from.

- The Skullsite website (skullsite.com) is a bird skull collection (1,600 species and growing) with photos and measurements.

Other books include those which are largely designed for bird ringing studies, and these include helpful insights into feather plumages, particularly when comparing cryptic species such as willow warbler and chiffchaff, and where only parts of the wing have been found:

- Baker, J. *Identification of European Non-Passerines*, 2nd edition. BTO, Thetford, 2016.

- Demongin, L. *Identification Guide to Birds in the Hand.* Beauregard-Vendon, 2016.

- Jenni, L. and Winkler, R. *Moult and Ageing of European Passerines*, 2nd edition. Bloomsbury, London, 2020.

- Svensson, L. *Identification Guide to European Passerines*, 4th edition. BTO, Thetford, 1992.

- Monronval, J.B. *Guide to the Age and Sex of European Ducks.* Office National de la Chasse et de la Faune Sauvage, Paris, 2016. Available online as a downloadable PDF in English through the IUCN Species Survival Commission Duck Specialist Group (ducksg.org) and as the original French version through ONCFS (oncfs.gouv.fr). Can also be purchased as a hard copy.

For mammals the following books are useful, particularly for illustrations of skulls:

- Barn Owl Trust. *Barn Owl Conservation Handbook.* Pelagic Publishing, Exeter, 2012.

- Brown, R.W., Lawrence, M.J. and Pope, J. *Animals Tracks, Trails and Signs* (Hamlyn Guide). Bounty Books, London, 2004.

- Macdonald, D. and Barrett, P. *Mammals of Britain and Europe* (Collins Field Guide). Collins, London, 2005.

The following book is also an essential guide for monitoring raptors and includes images of their own wing and tail feathers:

- Hardey, J., Crick, H., Wernham, C., Riley, H., Etheridge, B. and Thompson, D. *Raptors: a Field Guide for Surveys and Monitoring*, 2nd edition. Scottish Natural Heritage, Edinburgh, 2009.

Many museums have study skins and mounted skins of a range of wildlife, particularly birds and mammals. Contacting a museum curator and arranging a visit can be a useful way of identifying trickier feathers or skulls.

MY OWN STORY – LEARNING MY FEATHERS AND IDENTIFYING PEREGRINE PREY

As I child I loved collecting feathers, and I still have fond memories of where and how I found many of them. For example, I remember the moulted greenfinch feathers I found beneath a plum tree at primary school, the scavenged remains of a female sparrowhawk under a holly tree in woodland, the iridescent mandarin duck feathers moulted by a lakeside, the chewed remains of pheasants found on a country walk, and the feathers of lapwings and gulls on my school field.

Armed with my Dorling-Kindersley Eyewitness book of birds, which featured lots of wings and skulls of birds, I taught myself to identify feathers. Later I came upon the book *Tracks and Signs of Birds of Britain and Europe*, which had beautiful illustrations of feathers and skulls. It became my 'go to' book for anything new I needed to identify. I also loved simply flicking through the illustrations, taking in their colours, shapes and characteristics like a sponge. By the time I began identifying the prey of peregrines, in 1998, I had become familiar with most common birds and had built up my knowledge further to include species such as snipe, jack snipe and meadow pipit, all of which were taken by the peregrines.

I have now been studying the diet of urban peregrines for over 20 years. In the first few years there were several feathers that I was unable to identify. They were from a small bird, perhaps a large finch and no bigger than a starling and sandy-brown in colour with a white spot on the inner web. Back in 2000 there wasn't the wealth of websites and books that there is today. I started looking through the books for birds that had white on the inner web. The first bird I found was a hawfinch. I was convinced it was this species and even visited the Natural History Museum at Tring in Hertfordshire to compare it with the skins of hawfinches. But they didn't fit with the feather – the hawfinch skins showed wing feathers that were glossy and iridescent, and their shape and markings just didn't match.

A year or two later, with the feathers still unidentified, I remember looking through a book on waders or shorebirds with beautiful plates of illustrations. Suddenly, there, in great detail, was the bird that my feathers came from. It had all the right white spots in the right places. The bird in question was a common sandpiper, a species typical of a river habitat. With a little more research, I was able to confirm that they were indeed common sandpiper feathers.

On another occasion, when collecting peregrine prey remains in the city of Bath, I discovered the wing coverts of a corncrake, a rare bird in the UK, confined to the western isles of Scotland, the Nene Washes – where it has been reintroduced – and Northern Ireland. I spent a lot of time comparing the feathers with museum specimens, and it took many years before I was able to get it accepted by the local county recorders' team. Fortunately for me, other subsequent corncrakes were eaten by peregrines, providing more evidence and feathers to confirm the earlier identification. By studying the diet of peregrines, I have helped confirm that peregrines are occasionally taking corncrakes in the autumn over Exeter, Bath and Derby, no doubt as the corncrakes are migrating south at night.

Identifying feathers, skulls and other body parts isn't always this challenging, but it can be fun and is always very satisfying when you work out what they came from. Confirming an identification also provides important information for local county recorders: it tells us more about what species are doing and how they are using a local area – particularly for rare, shy or unusual animals that may otherwise go unnoticed by people.

Parts of a bird

When identifying the feathers you have found it may not always be clear where on the bird they are from. The longer, stiffer feathers may be from the wing or the tail, while softer and smaller feathers come from different parts of the main body. Throughout the book I refer to specific feathers, and the following photo diagrams highlight where on the body of a bird they are from. The most likely feathers to be found in prey remains are the primary and secondary wing feathers alongside the tail feathers. Softer, lighter back, rump, breast and belly feathers are most easily blown away, or may be found clumped on a lawn or along a woodland path.

Primary feathers   Primary coverts   Secondary feathers   Greater coverts   Tail feathers   Rump

Breast feathers Belly feathers Scapular feathers Flank feathers Greater coverts Tertial feathers Primary feathers

Tertial feathers

Secondary feathers

Greater coverts

Scapular feathers

Median coverts

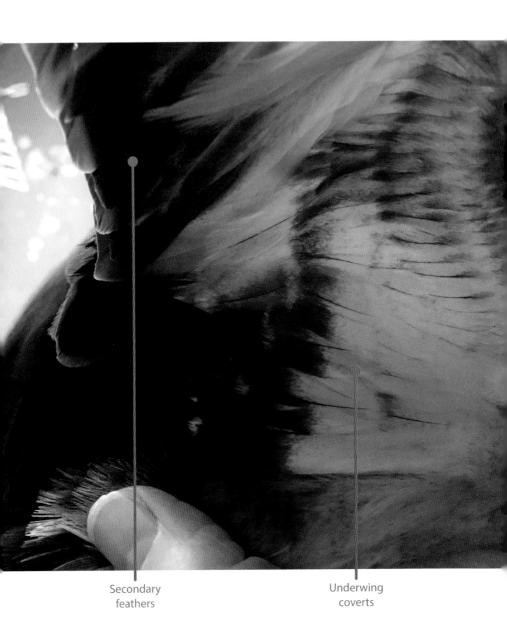

Secondary
feathers

Underwing
coverts

Glossary

Alula	The bird's thumb or pollex (first digit), a group of small feathers (usually three) on the leading edge of the wing that can stick out to form a small winglet, used to control flight particularly when slowing down and landing.
Barb	Parallel structures that branch off from the rachis. They fasten together using barbules to form the vane of the feather.
Barbule	Tiny hooked branches along each barb that fasten them together.
Barred	Feathers with many coloured bars running horizontally along the length of a feather.
Cache	A place where a bird of prey, such as a peregrine, stores excess food in a crevice in a cliff or on a building.
Coverts	A range of small feathers that overlay the primary, secondary and tertial feathers, and infill the rest of the wing above and below.
Cranium	The rear part of the skull that contains the brain.
Emargination	Where the outer web of a primary feather becomes narrower towards the tip. May extend some way down the feather.
Flanks	Feathers that border the belly of a bird and where the wing rests when closed.
Gonys	Part of the lower bill of a large gull that forms a bulge; in adult gulls such as herring gulls, it has a bright red spot.
Greater coverts	The largest of the coverts on the upperside of a bird's wing, covering the bases of the secondary feathers.
Inner web	The wider part of the feather vane that forms the inner edge along the rachis of wing and tail feathers.
Mottling	Mix of smears and spots of colour that form a marbled effect rather than defined markings.

Notch	Where the inner web of a primary feather becomes narrower towards the tip. May extend some way down the feather.
Outer web	The narrower part of the feather vane that forms the outer (leading) edge along the rachis of wing and tail feathers.
Primary feathers (primaries)	The longer outer wing feathers used in flight.
Rachis	The central stiff structure that runs down the middle of a feather. Also knows as the shaft.
Secondary feathers (secondaries)	The shorter inner wing feathers used in flight.
Scapular feathers	Feathers that overlay where the wing joins the body of a bird.
Speculum	The shiny, iridescent patch on the secondary feathers of many ducks.
Streamer	An elongated, tapered feather, for example the outer tail feather of a swallow.
Talon	The sharp claw of birds of prey or raptors.
Tertial feathers	The innermost flight feathers of the wing. Usually longer and wider than the secondary feathers they border.
Vane	The area of feather that forms a flat sheet of linked barbs on either side of the rachis, made up of the outer and inner webs.
Vermiculation	Thin wavy bands of colour running in horizontal or vertical lines along a feather, often very close together.

Mallard
Anas platyrhynchos

Mallards, like many other ducks, have a striking patch on the secondary feathers known as the speculum. These feathers have a shiny, iridescent blue-purple outer web that is edged black with a thick white tip. Male body feathers reveal a range of deep chestnut-brown breast feathers and white or grey vermiculated belly and flank feathers. Black central tail feathers with curly tips may also be found. Female body feathers are a mix of dark brown and buff-orange bands that radiate down the feather from the tip or below.

Right wing (female), showing iridescent secondary feathers (speculum)

Breast and belly feathers (female)

Secondary feathers (speculum) with a white/grey underwing covert

Breast feathers (female)

Primary section of wing (female)

Teal

Anas crecca

Male teal have striking black and white vermiculated feathers, while females have brown and white barred feathers. Male teal have grey, velvety tertial feathers with a black and a white line running down their length. Their undertail coverts are black with creamy-white tips. Primaries are short compared to a mallard's; as in all ducks, these have a shiny panel running down the length of the underside of both inner and outer webs.

The iridescent green speculum is one of the most distinctive features. The outer secondaries, close to the primaries, are black and may have just a small patch of green. In females, more secondaries are black.

Head (male) FH   Head (female) CR   Belly feathers (male) HS

Breast feathers (male)   Right wing (female) CS   Belly feathers (female) HS

Outer (top) and inner (bottom) secondary feathers

Tertial feather (male)

Greater covert

Tertial feathers (male)

Other ducks

Ducks are commonly eaten by raptors and other predators. Their large size often means they are plucked and eaten where they are grounded rather than carried away. Clues that help with the identification include the speculum (the iridescent patch on the secondary feathers) and the colour or pattern of other feathers, particularly if a male duck in breeding plumage has been eaten.

Pintail *Anas acuta*: pair of female wings MK

Wigeon *Mareca penelope*:
breast and underwing feathers

Mandarin duck *Aix galericulata*:
female wing and tail feathers

Eider *Somateria mollissima*: male wing

Red grouse
Lagopus lagopus

The primary feathers are raptor-like, with notches and emarginations. However, they are much shorter than anything they resemble such as a buzzard or eagle, with uniform brown webs and off-white bases. Their tips have sandy-cream squiggly markings which extend right across the outer webs of the secondaries. The tail feathers are dark brown and may have some sandy-orange squiggles and marks at their tips. Legs and toes are feathered with long curving claws and may resemble those of a rabbit or hare.

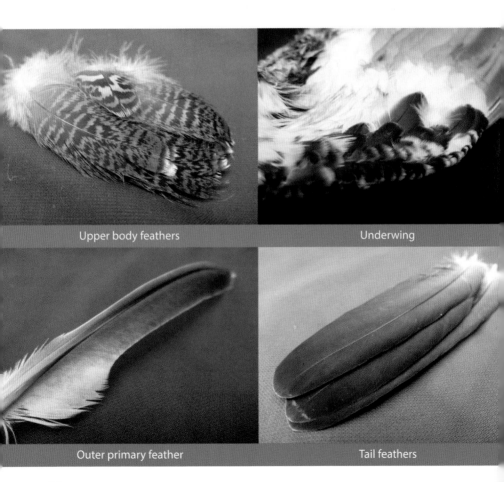

Upper body feathers

Underwing

Outer primary feather

Tail feathers

Carcass SH

Outer wing feathers

RED GROUSE 27

Feathered foot in heather

Inner primary feathers

Red-legged partridge
Alectoris rufa

RM

The most distinctive and easily spotted feathers are those from the flanks; they have a pale blue centre and adjoining strips that are cream, black and orange-brown in colour. The primaries and secondaries are dark brown with creamy-buff bands running along their outer webs. Tail feathers are rufous-brown. Most likely to be found in open country and alongside hedgerows or stone walls.

Flank feather

Neck/breast feathers

Upper body feather

Primary feathers

Flank feathers

Secondary feathers

Plucked body and tail feathers

Flank feathers

Buzzard chicks with eaten carcasses of red-legged partridges on edge of nest AF

Grey partridge
Perdix perdix

FVa

The upper body feathers have cream-yellow lines along their shafts and black and sandy squiggles or vermiculations. An adult head is orange, contrasting with a grey, finely vermiculated neck and breast. If male, the dark orange-brown feathers that form a horseshoe shape on the belly may be found. The flank feathers are grey and vermiculated with thick orange-brown bands. In younger birds the breast and flank feathers are orange-brown with cream streaks running along their shafts.

Carcass of a tagged bird RB

Outer primary feathers SC

Upper body feathers SC

Wing and feather remains of a tagged bird RB

Primary feathers
(bottom) and
flank feathers
(middle) RB

Feather remains RB

Quail

Coturnix coturnix

Quail have wing feathers no bigger than those of a starling. The primaries and secondaries are sandy-brown with cream-yellow bars and squiggles. The outermost primary is strongly curved, lacks any distinct bars and has a cream-yellow edge running along its length. The upper body feathers, like those of a grey partridge, have distinctive thick cream-yellow lines running along their shafts. The outer and inner webs, however, differ from the partridge's, alternating between cream-yellow, black and orange-brown markings of varying thickness.

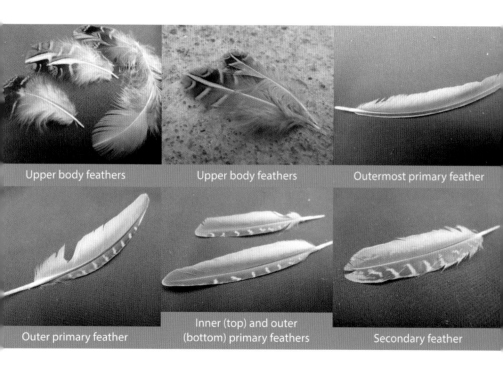

Upper body feathers

Upper body feathers

Outermost primary feather

Outer primary feather

Inner (top) and outer (bottom) primary feathers

Secondary feather

Body with legs

Right wing

Pheasant

Phasianus colchicus

RH

Pheasant feathers come in a whole range of patterns and colours. Those of the female combine sandy-cream backgrounds with a huge variety of brown markings; the tail is marked with alternating bars of dark brown, orange-brown and creamy colours. Male pheasants sport a variety of orange, brown, green and black plumage. The most distinctive feathers are those from the flanks and edges of the breast. They are bright shiny orange with pink-purple iridescence, black tips and grey-brown fluffy bases. Their tail feathers are long and tapering with a yellow-grey background, dark bars and loose orange-coppery edges with a pink-purple iridescence.

Body feathers (female)

Body feathers (female)

Secondary feather (female)

Body feathers (female)

Tail feathers (male)

Tail feathers (female)

Flank/breast
feathers (male)

Flank/breast
feathers (male)

Tail feathers (male)

Outer tail feather (female)

Body feathers showing green iridescence (male)

Plucked body feathers (male)

Chicken

Gallus gallus domesticus

Chicken feathers, alongside those from pheasant, are some of the most frequently found prey remains both in woodlands and in gardens. Chickens come in a huge variety of shapes and sizes, and their feathers reflect this. Their beautiful colours and patterns often lead to several different ideas of what they might be from. The wing feathers are narrow and strongly curved, and tail feathers can be elaborate, broad and colourful. The remains of chickens commonly include those with black and grey stripes or those that are pale orange. If you are in doubt about an unidentified feather, it is often from chicken or pheasant.

Tail feathers OK   Body feather SC   Leg MS

Body and wing feathers SB

Mix of feathers JH

Storm petrel

Hydrobates pelagicus

Storm petrels have primaries that at first glance look like those of swallows and house martins. The tail feathers are large and broad for such a small bird. The secondaries are matt black with white at the base of their inner webs. Remains may be found on seabird islands, along the coast and inland after storms. Smelling the feathers reveals the distinctive musty smell of a petrel.

Secondary section of wing

Secondary feather

Primary feather

Tail feather

Manx shearwater
Puffinus puffinus

RM

Manx shearwaters are often taken by gulls and raptors on islands, and can also be found dead or predated inland after stormy weather. The wings are black-brown on the upperside; on the underside they have contrasting white coverts. The primary section is thin and tapering, projecting a long way beyond the secondary feathers when closed. The feathers have a fishy, musty smell and their upper surface can look finely dusted, an effect created by salt from sea water.

Section of wing showing greater coverts and base of primaries

Primary section of wing

Underside of wing

Upperside of wing

Tail feather

Outer primary feather

Inner wing/mantle feather

Secondary feather

Underside of wing

Middle primary feather

Manx shearwater carcass (top) and feral pigeon feathers (bottom left) LS

Wings, bones and skull LS

Little grebe
Tachybaptus ruficollis

The legs are flattened with obvious large scales; they are blue-grey and fade to grey-brown or black when the bird is dead. The toes are lobed – each toe has a flange of skin on either side that folds back when the grebe is swimming. The belly and breast feathers are strongly curved and loosely structured with a dark shaft; the barbs and barbules are grey or white. The grey-brown wing feathers look too short for a bird of this size. They are curved, and the outer primaries are strongly notched. The secondaries have white on the inner webs and may be tipped white; they lack the full white on the outer webs of other grebes.

Lobed foot FH

Lobed foot FH

Breast feathers

Left wing

Inner wing

Secondary feather

Dismembered carcass on top of high-rise office

Carcass FH

Lower body, wing and legs FH

Sparrowhawk
Accipiter nisus

Sparrowhawk feathers are distinctly barred; the outer primaries are strongly emarginated and notched; the male's wing feathers are jackdaw-size or smaller while those of the female are more carrion-crow-size, both shorter and narrower than those of buzzard. The tail feathers are long and narrow. If the talons are found, they are long, very thin and delicate-looking, while those of buzzard and peregrine are thicker and more robust. Those of kestrel are shorter and slightly thicker.

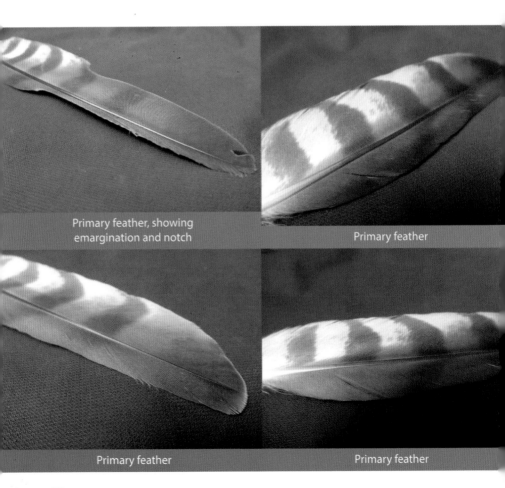

Primary feather, showing emargination and notch

Primary feather

Primary feather

Primary feather

Wings (underside) JHu

Wings above JHu

Juvenile peregrine in Norwich with sparrowhawk prey HOT

Water rail

Rallus aquaticus

The most obvious feathers found are the stripy flank feathers, which have a loose structure and are often tipped yellow. Accompanying these are blue-grey breast and belly feathers. The primaries and secondaries are shorter, narrower, and often darker than those of moorhen. The tips have a subtle frayed fringe. The heads are often found, revealing the gently curved blood-red beak of adults, although this is darker and less red in younger birds. The head feathers are blue-grey. The toes are long; the legs and toes are pink-beige or grey in colour.

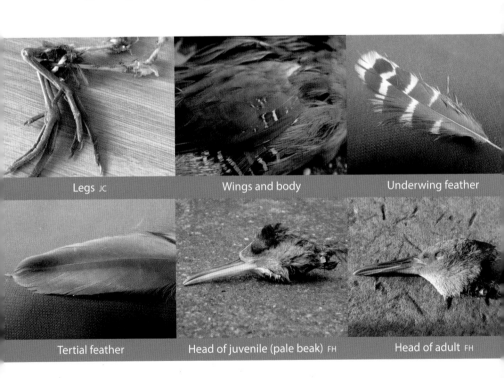

Legs JC

Wings and body

Underwing feather

Tertial feather

Head of juvenile (pale beak) FH

Head of adult FH

Pair of wings (at barn owl roost) SB

Half-eaten carcass CS

Corncrake

Crex crex

Corncrakes, while secretive by day, migrate at night and are taken by raptors, particularly urban-dwelling peregrines. Their wing feathers have a similar shape to those of a water rail or moorhen; there is a subtle fringe to the edge of the primaries and primary coverts. The immediate difference is the orange-brown colour of the corncrake's feathers. Other feathers found might include the breast feathers, which are white with faint orange-brown lines.

Left to right: Secondary feather, two primary coverts and covert

Flank feather

Wing showing tertial feathers

Wing showing coverts

Tertial feather

Outer primary feather

Secondary feather

Primary covert

Primary feather showing loose fringe

Wing covert

Spotted crake

Porzana porzana

The spotted crake is a nocturnal migrant, appearing in the diet of urban peregrines during autumn migration. Their wing feathers are similar to those of water rail, but smaller. Their primaries also have the thin area of frayed edging on the very tips. The clinching identification features are the white, lightning-shaped marks on the wing coverts and tertial feathers.

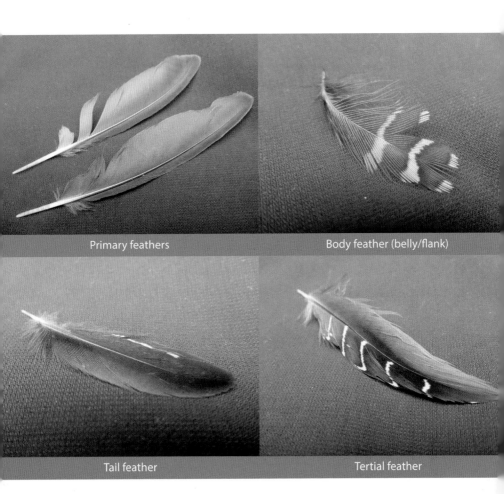

Primary feathers

Body feather (belly/flank)

Tail feather

Tertial feather

Mix of inner wing feathers WL

Secondary feather

Wing covert

Moorhen
Gallinula chloropus

The lower body feathers of a moorhen have a loose structure and a light grey appearance, particularly on the belly. The wing feathers are dark brown-black and lack any emarginations or notches. The flank feathers have white lines running along their length. These are coffee-coloured in younger birds. Undertail coverts are white. The tail feathers are short and dark brown, almost black; their undersides are paler brown with a black tip. The legs of adult moorhens are yellow-green, with red towards the top of the leg in more mature birds. Those of younger birds are brown-green. If an adult has been dead for a while, the bright colours in its legs will deteriorate to become dark green or black.

Pair of wings

Flank feathers

Head (juvenile) SC

Lower mandible JB

Wings and legs (juvenile) CR

Outer primary (underside)

Upper mandible FH

Mix of breast and belly feathers

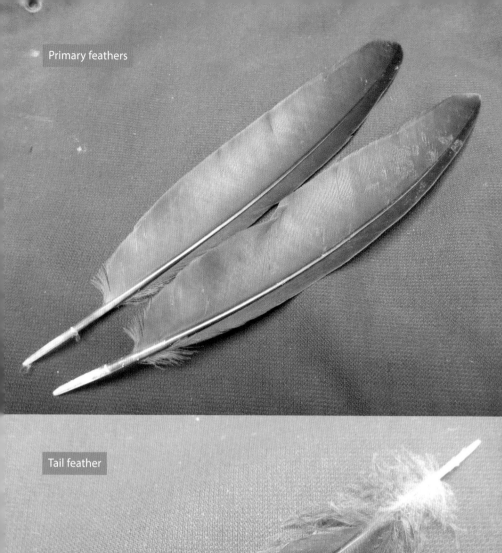

Primary feathers

Tail feather

Coot

Fulica atra

The body feathers of a coot have a loose structure and a uniform grey-black appearance. The primaries are broader than a moorhen's and dark grey with paler inner webs. The secondaries are similar, except that the tapering tips become very light grey or white. The legs are large, silver-grey and with hints of yellow and red in older birds. The toes are long and lobed, each toe with its own rounded flap of skin or lobe.

Live bird, showing lobed feet   Leg   Breast feathers

Skull   Sun-bleached primary feathers

Oystercatcher
Haematopus ostralegus

Oystercatchers have wing and feathers not dissimilar to those of the black-tailed godwit. In the oystercatcher they are larger, and are sometimes accompanied by the distinctive skull with a long orange beak, and thick pink legs. White bars on the outer webs of the middle and inner primaries are topped and tailed with black, while in the godwit the white bars on the outer webs are restricted to the basal half without any black beneath them.

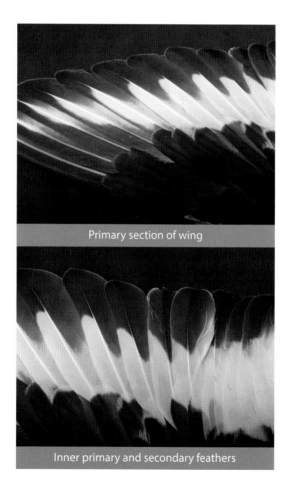

Primary section of wing

Inner primary and secondary feathers

Outer primary feather

Inner primary feather

Beak

Outer primary feathers

Inner wing (juvenile)

Avocet
Recurvirostra avosetta

Avocets are relatively easy to identify, depending on which parts of the body are found; the long blue-grey legs and decurved beak are distinctive. The outer primaries are black apart from a small amount of white at the base of the inner webs. This white becomes more extensive further along the wing until the two inner primaries and secondaries are all-white. If wings are found, the white of the secondaries, greater coverts, alula and lower sections of the primary coverts contrasts with the black of the scapular feathers, median wing coverts and outer primaries.

Pair of wings FH

Leg FH

Wing with sandy-buff juvenile feathers (black in adult)

Secondary feather

Primary feather

Eaten head remains FH

Lower mandible

Lapwing
Vanellus vanellus

The most obvious feathers are the dark green iridescent back and scapular feathers and the long orange undertail coverts. The outermost black primaries have creamy-white patches that appear as distinctive elongated smudges towards the tips. The tail feathers are square-ended, white with thick black tips (often edged white) that extend a third of the way down. The secondaries are mostly black with a green sheen on the outer web and a contrasting white segment towards the base. The legs are dark red when fresh and fade to grey-brown, while those of golden plover are black, slightly smaller and thinner-toed.

Upper body feather

Carcass JB

Undertail covert

Outer primary feathers

Pair of wings and head ML

Secondary feather

Tertial feather

Primary feathers

Merlin with lapwing GG

Pair of wings MK

Golden plover
Pluvialis apricaria

GT

Golden plovers have unmistakable yellow-spotted feathers covering the back, scapular feathers and wing coverts. The spots are bright lemon-yellow, particularly in fresh feathers. On older, worn feathers the yellow will be faded. The primaries and secondaries are a combination of light chocolate-brown and white markings. The thin outer primaries have a white portion on the outer web, topped and tailed with brown. The legs are black and often lightly soiled with mud.

Scapular feathers

Mix of primary (left), secondary and covert feathers

Central tail feather JB

Head FH

Primary feathers SC

Tail feather and upper mandible

Outer primary feather, showing the white portion of the shaft

Carcass revealing wings, legs and scapular feathers AL

Grey plover
Pluvialis squatarola

Grey plover wing feathers are very similar to those of the golden plover; however, they are monochromatic and lack the golden-yellow hues, including the spotted wing coverts. The innermost underwing coverts (the axillaries) are black, unlike those of golden plovers and other waders.

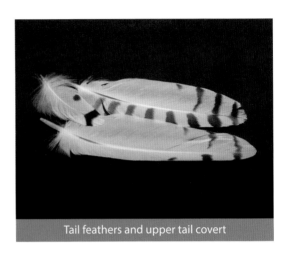

Tail feathers and upper tail covert

TAIL FEATHERS: GOLDEN PLOVER OR GREY PLOVER?

The tail feathers of these two species are similar in size. The left-hand feather is from golden plover, showing dark black bands separated by thin pale yellow bands. The grey plover feather on the right lacks the yellow and dark brown barring; in this species the tail feathers are white with varying coverage of dark brown bars, and very few on the outer tail feathers.

Secondary feather and wing coverts

Secondary feather

Upper body feather

Tail feathers

Ringed plover
Charadrius hiaticula

Ringed plover wing feathers are very similar in size, colour and pattern to those of dunlin. A dunlin has a thinner white edge running down the outer web of its primaries; in the ringed plover this is a thicker wedge of white extending to the shaft. The shafts of the primaries are dark with a white middle area that forms part of the white wedge in the inner primaries. Ringed plovers have striking tail feathers with white tips and dark bands. Their greater coverts are sandy-brown and lack the white tips of dunlin.

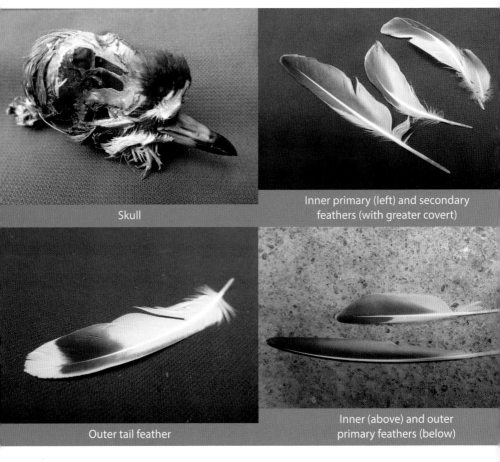

Skull

Inner primary (left) and secondary feathers (with greater covert)

Outer tail feather

Inner (above) and outer primary feathers (below)

Wing revealing tertial and primary feathers

Tail feathers

Whimbrel
Numenius phaeopus

RV

Whimbrel have dark brown wing feathers with obvious cream-white notches on the inner and outer webs. Similar patterns are found on the greater coverts and across the rest of the wing. The tail feathers are barred – the bars are curved or kinked against a light brown background. Although similar, the curlew has a white background to its tail feathers and the wing feathers are both longer and lighter brown. Location is a helpful clue – if remains are found in an urban location they are most likely to be from a whimbrel. Here whimbrels are regularly caught by peregrines as they migrate overland.

Tail feather

Primary feather

Upper body feathers JMa

Secondary feathers

Underwing

Primary and secondary sections
of wing showing coverts

Underwing feather

Bend of wing, showing coverts

Inner section of wing, showing tertials and coverts

Half-eaten carcass in a car park (peregrine kill) SH

Secondary feathers of curlew (above) and whimbrel (below)

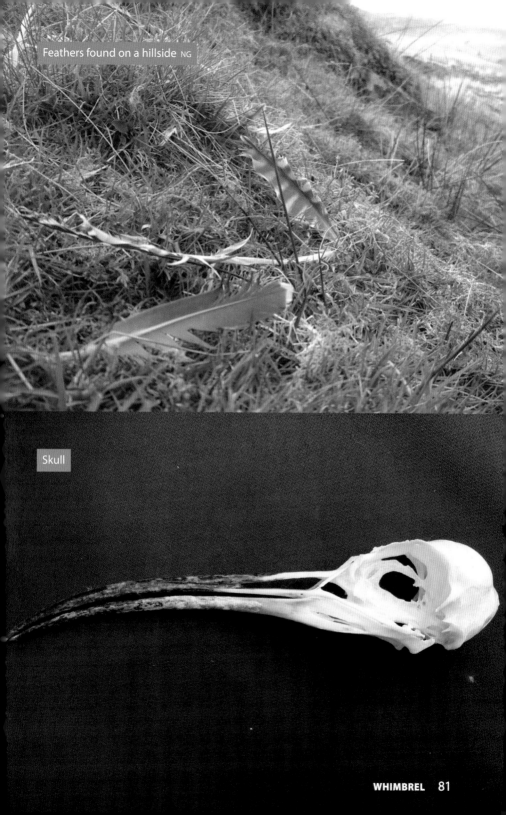

Feathers found on a hillside NG

Skull

Curlew
Numenius arquata

Wing feathers are longer than those of whimbrel. Secondaries have bigger, bolder white markings; these are more diffuse in whimbrel. Feathers of the curlew are lighter and more sandy in colour than those of whimbrel, particularly during the summer months when they fade with sunlight. The beak is much longer and more gently curving; whimbrel has a shorter beak that bends more abruptly.

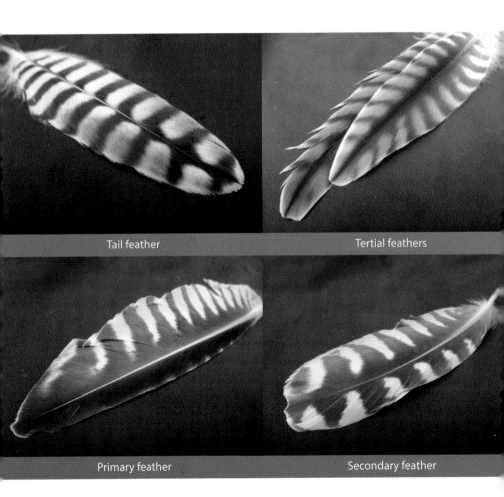

Tail feather

Tertial feathers

Primary feather

Secondary feather

Bar-tailed godwit

Limosa lapponica

Bar-tailed godwits lack the striking white and black contrasts in the wing and tail feathers that are seen in the black-tailed godwit. Instead, they have mottled brown and white feathers. The primaries have a white shaft and a dark brown outer web and tip. The inner web of these feathers has a dark section running down its length, diffusing into mottling and then white. The inner primaries and secondaries are uniform sandy-brown with a white-edged outline. The tail feathers are barred white and sandy-brown. In summer plumage, deep orange belly/breast feathers may be found alongside orange and dark brown wing coverts.

Inner wing

Primary feather

Orange lower body feather
(summer plumage)

Left to right: two primaries,
two tertials, tail feather

Wing covert

Tertial feather

Wing covert

Outer tail feather

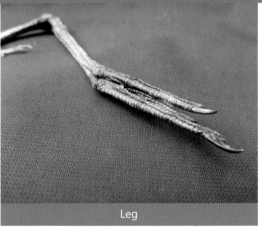

Leg

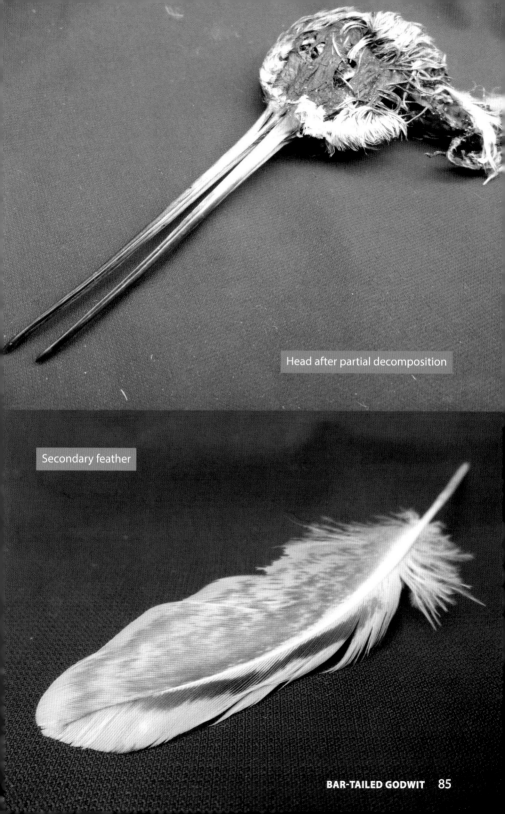

Head after partial decomposition

Secondary feather

Black-tailed godwit
Limosa limosa

Black-tailed godwits have striking black and white feathers. The shaft of the primaries is white. Approximately a one-third section of the primaries and secondaries, towards the tip, is black or black-brown with a concave curve on the inside web. The tail feathers are up to two-thirds black, often with a thin subtle sandy-brown tip. Inner wing feathers may be sandy-brown, particularly on older faded feathers. Those predated when in breeding plumage reveal a variety of tanned-orange and black feathers, often with bars or splotches.

Inner primary feather

Secondary feather

Tail feather

Left to right: belly feather, secondary, three tail feathers SC

Skull

Summer plumage breast feather on a lawn beneath a peregrine roost site sc

Upper wing coverts (summer)

Secondary (top) and tertial (bottom) feathers WL

Greater covert

Knot

Calidris canutus

Knot wing feathers can be difficult to identify if only the outer primaries are found, as they resemble those of turnstone and ruff. Look for a range of clues including black and white V-shaped patterns on the flank feathers and pale grey tail feathers with tips thinly edged white and wavy patterns on their inner webs. The middle and inner primaries have narrow white edges, also seen in the much smaller dunlin.

Head FH

Underside of primary section of wing

Primary feathers, showing white shaft

White leading edge on outer webs of middle and inner primary feathers

Dunlin
Calidris alpina

RM

Dunlin wing feathers are narrower and shorter than a starling's. The primaries stand out amongst prey remains for their white shafts, dark tips and outer webs with white edges – these white edges become increasingly thick towards the inner primaries. Secondaries are dark grey-brown and white across their basal half. Tertial feathers are much longer than the secondaries, with varying patterns depending on age and season. The long greater coverts are dark brown with large white tips. The short legs and feet are black.

Tertial feather (juvenile)

Secondary section of wing

Left wing BL

Upper body feather (left) and tail feather (right)

Primary feathers and legs

Left wing

Woodcock
Scolopax rusticola

GT

The head and beak are thicker than in a snipe, although side by side the beak may be a similar length. The rusty-orange and brown barred wings feathers are stiff and slightly curved. At first glance the wing feathers, particularly the secondaries, might be confused with a kestrel. Kestrel primaries have a dark outer web and, along with the secondaries, inner webs with bold white-orange patches. Woodcock tail feathers are short and wide, with orange and dark brown bars. They have distinctive tips that are matt grey on top and bright silvery-white beneath.

Head

Left to right: breast feather, primary, three tertials, secondary, two primaries SC

Tip of tail feather (underside)

Breast/belly feathers

Undertail covert

Carcass CS

Left wing AL

Jack snipe
Lymnocryptes minimus

While at first the remains of jack snipe and snipe may look similar, there are some distinctive feathers that help tell the difference between the two. In jack snipe the secondaries are distinctly tapered, forming a pointed tip. Their tail feathers are short and very uniform compared to snipe. The upper body feathers that run along the back have a thick creamy-blonde streak, adjacent to a streak of mixed black and orange-brown marks. The edges of these feathers have a thick band of green iridescence running along their length. If the head, skull or beak is found the beak itself is shorter than that of either snipe and woodcock, with a thick base.

Head FH

Primary feather

Secondary feather

Tail feather

Upper body feather

Secondary feather

Snipe
Gallinago gallinago

Snipe are frequently taken by several raptor species, and have a huge range of cryptic and striking feathers. The most obvious remains are the heads and beaks; the cranium is often broken and damaged. The feathers most likely to be found are the plain grey-brown primaries (with paler inner webs), white-tipped secondaries and striking tail feathers. Small, black and white stripy underwing feathers may also be found. The legs and long toes are olive-green.

Secondary feather

Head JB

Mix of wing and body feathers

Primary covert (top), breast and flank feathers

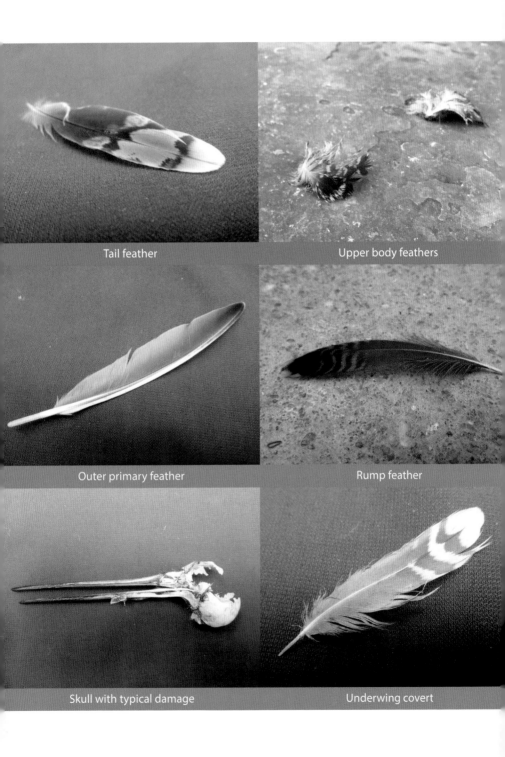

Tail feather

Upper body feathers

Outer primary feather

Rump feather

Skull with typical damage

Underwing covert

Buzzard carrying snipe CH

Beak

Common sandpiper
Actitis hypoleucos

RM

Wing feathers are small, barely more than chaffinch or sparrow size. Distinctive white wedges are present on the inner webs of the primaries. These white areas make up half the lower portion of the secondaries and extend onto their outer webs too. Hawfinches have similar white wedges on the primaries, although the white is more extensive on those closer to the secondaries (which lack the white) and accompanied by a purple iridescence on the outer webs. Common sandpiper legs and toes are yellow-brown. Their wing coverts are grey-brown with fine mottling.

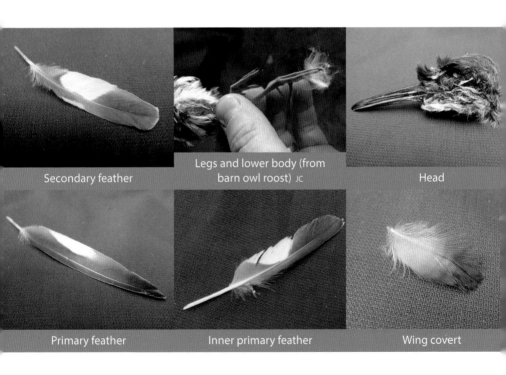

Secondary feather

Legs and lower body (from barn owl roost) JC

Head

Primary feather

Inner primary feather

Wing covert

Green sandpiper
Tringa ochropus

The primary and secondary feathers of green sandpiper are dark brown, almost black, with a subtle greenish sheen. The tail feathers are a mix of white or black and white, and the legs and toes are olive-green (like snipe). Look for very fine pale spotting against a dark background on the upper body feathers and wing coverts. The pale or white spots on many feathers, such as the scapulars, are found along the outer edges of the outer and inner web.

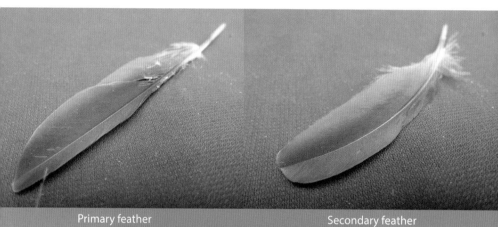

Primary feather

Secondary feather

Tail feathers

Female peregrine with the remains of a green sandpiper: contrasting white underparts and dark, uniform wings, barred underwing coverts and green legs FH

WAKEFIELD CATHEDRAL
WAKEFIELD NATURALISTS' SOCIETY
wakefieldnaturalists.org

NestingBox

Plucked remains, revealing spotted wing coverts, barred underwing feathers and green legs (hen harrier nest) GC

Redshank

Tringa totanus

Redshank feathers are very distinctive, with very white inner primaries and secondaries with dark squiggly markings. The outer primaries look blacker than those of other wading birds. If present, the heavily barred tail feathers, patterned tertial feathers and coverts, and red legs all help confirm identification. The beak is long and tweezer-like, lacking the rounder, thicker tip of longer-beaked birds such as godwits.

Secondary feathers

Primary feather

Wing coverts

Tail feathers

Beak

Tertial feather

Mix of feathers

Flank feather

Primary feather

Fresh carcass (sparrowhawk kill) MJ

Upper body feathers (summer)

Black-headed gull
Croicocephalus ridibundus

Adult black-headed gulls have bright red or orange-red legs, white tail feathers, and silver-grey secondaries, inner primaries (with some blackish areas on tips) and wing coverts. The outer primaries are white with black outer webs and grey-white tips against a black background.

Juvenile black-headed gulls show a dark brown band on the tips of the tail feathers. The secondaries and wing coverts have brown across their inner and outer webs. The black on the primaries is more extensive. Legs and feet are pale orange.

Secondary feather (juvenile)

Inner primary feathers (adult)

Wing coverts (juvenile)

Tail feather (juvenile)

Mix of tail and body feathers GJ

Legs AL

Head (adult)

Cached black-headed gull (adult) JB

Pair of wings (adult) AL

Peregrine with a juvenile black-headed gull CL

Gulls (chicks and fledglings)

In towns and cities, the chicks and fledglings of lesser black-backed gulls *Larus fuscus* and herring gulls *L. argentatus* are frequently run over by cars or taken by raptors after they become grounded. Their remains are often found when they have very short wings and their feathers are just growing. The primaries and secondaries are black-brown while the wing coverts are a mix of creamy-white, brown and black. The tail feathers are a striking mix of these three colours. Chick feathers of both species are very similar; herring gull chicks have paler/grey inner primary feathers.

All the photos on this page are of lesser black-backed gull.

Mix of feathers

Secondary/tertial feathers

Tail (bottom), secondary (middle) and covert (top)

Primary feathers

Left wing (above) JM

Left wing (below) JM

Right wing of lesser black-backed gull chick at 3–4 weeks old, showing uniform dark inner primary feathers

Right wing of herring gull chick at 3–4 weeks old, showing lighter mottling on inner primary feathers

Right wing of lesser black-backed gull chick, showing the tertial feathers' dark centres and continuous creamy outline

Right wing of herring gull chick, showing tertial feathers' broken creamy outline. The rest of each feather is lighter, with further markings running across the width of the feather

Adult gulls

Several gull species may be taken by raptors around coastlines and in towns and cities where they breed.

Herring gulls have pink legs, grey inner primaries and secondaries with white tips. The outer primaries have black tips with distinctive white spots, known as mirrors, and variable amounts of black and grey along their outer webs.

Lesser black-backed gulls have feathers of a similar size to those of the herring gull. However, the outer primaries are sooty-grey with black tips containing white spots (mirrors), and the inner primaries and secondaries are sooty-grey with white tips. They have yellow legs.

The wing feathers of kittiwakes *Rissa tridactyla* are slightly shorter and narrower than those of the herring gull. They also lack the mirrors and the black outer webs. Instead, their outer primaries look as if they have been briefly dipped into black paint. If heads are found, the herring gull has a thick, robust beak with a red patch on the lower mandible (the gonys); the kittiwake has a thinner yellow beak, and in the summer a bright red mouth.

Great black-backed gull *Larus marinus* feathers are much larger than both the herring and lesser black-backed gulls, with thicker white tips and spots. Common gull *L. canus* feathers are similar to those of kittiwake, but with more extensive black and white spots.

Young gulls in their first two or three years of life are more variable in colour and pattern, based on a pallet of brown, black, grey and white wing and tail feathers.

Lesser black-backed gull   Kittiwake   Herring gull

Herring gull wings LS

Kittiwake carcass LS

Kittiwake wing

Herring gull primary feathers LS

Kittiwake: tips look as if they've been dipped in black paint. No white spots on outer primary feathers LS

Sandwich tern

Thalasseus sandvicensis

Primary feathers of Sandwich tern have thick margins of white along the inner web and the tip. Compared with adults, juvenile wing feathers are darker, with the middle to outer primaries having a blacker band running down the inner web alongside the shaft. The leg and feet are black, and the beak of adults is black with a yellow tip.

Primary feather (juvenile)

Primary feather (adult, fresh)

Secondary feather (adult, fresh)

Wing

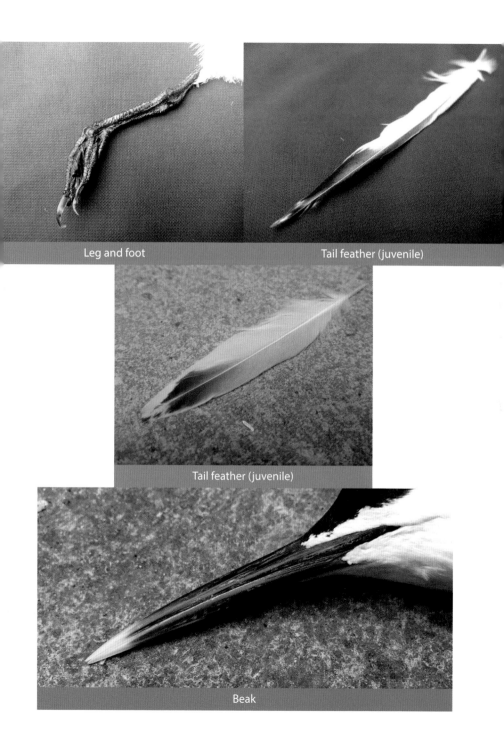

Leg and foot

Tail feather (juvenile)

Tail feather (juvenile)

Beak

Primary feathers

Outer primary feather

Common tern

Sterna hirundo

Remains of common and Arctic terns *Sterna paradisaea* are almost identical. If available, however, the outer primary feathers confirm the species. If the black-grey margin on the inside web is more than 5 mm wide it is from a common tern. If it is less than 5 mm it is from an Arctic tern. In both species, the secondaries are strongly curved, while the outer primaries are long, narrow and pointed. The inner primaries are considerably shorter. The outer webs of the primaries and primary coverts are covered in a white-silver powder-like layer.

Left to right: primary, two secondaries, tail feather, secondary

Secondary feather

Leg and foot

Primary feather, showing thick black-grey margin on inner web

Pair of wings AL

Upper mandible and forehead FH

Auks

Whole wings and bodies are often found, and if the heads or skulls are present their identification is an easier process. If the head is missing, puffins *Fratercula arctica* and razorbills *Alca torda* have black wings and upper bodies while those of the guillemot *Uria aalge* are brown-black. Puffins appear smaller and more compact than razorbills and guillemots. All have short wings; the primaries are narrow, curved and pointed, while the secondaries are short and curved with white tips. The black guillemot *Cepphus grylle* in the breeding season is a striking bird with black plumage, white wing patches and bright red legs and mouth.

Razorbill

Puffin

Guillemot

Black guillemot

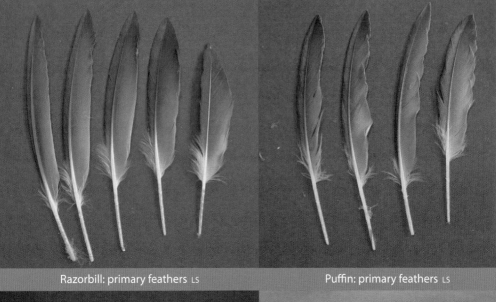

Razorbill: primary feathers LS

Puffin: primary feathers LS

Razorbill: secondary feathers LS

Puffin: primary wing section

Guillemot: carcass

Black guillemot: carcass
(with leg ring) GG

Feral pigeon
Columba livia

ARo

The remains of feral pigeons are among the prey items most commonly encountered with raptors such as *Accipiter* hawks and peregrines. Feral pigeons come in many colours including white, grey, blue-grey, orange-red and black. A whole range of feathers and body parts may be found. Their legs are dark red in adults or pink-grey in young birds. Tail feathers are half to two-thirds the size of those from a woodpigeon. The wings are pointed and falcon-like although they lack the alternating dark and light bars on the inner web. One or both wings are often found attached to the breastbone.

Primary section of right wing

Mix of primary feathers from different birds

Underwing (red variety) CS

Primary section of right wing AL

Inner secondary feathers (red variety) AL

Primary section of wing and adjoining bones CS

Wing coverts AL

Iridescent neck feathers AL

Tail feather (dark) AL

Tail feather AL

Old head AL

Leg and adjoining bones AL

Stock dove
Columba oenas

RM

The stock dove's feathers are similar to those of the feral pigeon. If the wing feathers are found, the middle and inner primaries have a blue-grey lower portion on the outer web; the rest of the feather is black, lacking the uniform grey of feral pigeons and the white outline of woodpigeons. Tail feathers are similar to those of feral pigeon. If the head is found, the beak is yellow like a woodpigeon's rather than black like a feral pigeon's. The tertials have distinctive blocks of black across their outer webs – which in the feral pigeon is more extensive and present on the secondaries too.

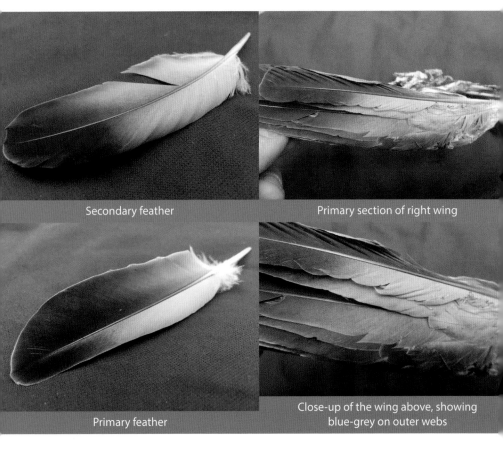

Secondary feather

Primary section of right wing

Primary feather

Close-up of the wing above, showing blue-grey on outer webs

Woodpigeon
Columba palumbus

Predated woodpigeon remains are commonly found in woodlands, open countryside, copses and gardens. Woodpigeons have distinctive telltale feathers. For example, the tail feathers have three thick bands which are most obvious on the underside. At the tip there is a black band, followed by a paler middle band (pale grey on the top of the feather, white-grey on the underside), and a darker band of grey closest to the base of the feather. The primaries are wider than those of a feral pigeon with a bright white line running down the outer web. Although feral pigeons may have a pale line, it is never as white. The secondaries are longer than those of a feral pigeon and less varied in colour, lacking any black bars. The body feathers of pigeons and doves are less easy to distinguish, and are best identified along with associated wing and/or tail feathers.

Mass of body feathers

Primary feather and primary covert (foreground)

Neck and upper breast

Tail feathers (foreground)

Tail feather (above)

Tail feather (below)

Primary feather

Damp primary feathers and coverts

Feather remains in woodland (predated by goshawk)

Turtle dove

Streptopelia turtur

At first glance turtle dove feathers look like those of a collared dove. A closer examination reveals that the wing feathers are generally shorter and narrower, with primaries sandy or dark brown with a less distinct pale edge. In juveniles this edge is pale orange-brown. The tail feathers are shorter and more contrasting, above and below, with whiter tips featuring on the upperside. Their diagnostic orange wing coverts with dark centres are often found too.

Tail feather

Tail feather

Wing coverts

Outer primary

Inner primary

Primaries (juvenile)

Secondaries

Mix of wing and tail feathers

Sun-bleached primaries

Collared dove
Streptopelia decaocto

Collared doves have dark primary feathers edged white and with lighter grey bases. The outer primaries have clear emarginations and notches. The tail feathers are distinctive – the central pair are a buff-grey above and below, while the other tail feathers have a white upper half and a darker lower portion. The amount of white lessens towards the outer edge of the tail. The collared dove's buff-grey body feathers are often found, helping with identification.

Carcass CS

Primary section of wing JB

Tail feathers (central tail feather, bottom)

Outer tail feather

Cuckoo
Cuculus canorus

Cuckoo feathers come in a whole range of different colours and markings depending on which part of the body they are from and the age and sex of the bird. Juvenile and rufous female cuckoo primaries resemble those of kestrels – look for the curved shape in the cuckoo and the lack of notches in the inner web of outer primaries. Identification is helped by other feathers such as the long, loose rump feathers and multi-patterned long and broad tail feathers.

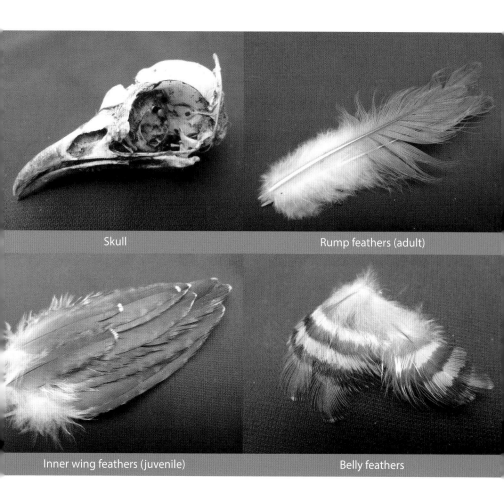

Skull

Rump feathers (adult)

Inner wing feathers (juvenile)

Belly feathers

Inner primary feather (adult)

Undertail covert (adult)

Primary feathers (adult)

Primary feathers: juvenile
(left) and adult (right)

Outer primary feather (adult)

Tail feathers: adult (left) and juvenile (right)

Outer tail feather (adult)

Plucked wing from a peregrine kill, with most feathers moulted into adult plumage but still showing one juvenile greater covert and one alula feather (sandy-brown with orange-cream markings) MPi

Underside of plucked wing showing distinctive white bars on inner webs and fine barring on coverts around bend of wing MPi

Barn owl

Tyto alba

RM

Barn owl remains are distinctly white, especially from a distance. Close up, the wing, tail and body feathers reveal shades of yellow-orange alongside faint or strong brown bars. As in other owls, the surfaces of the feathers look and feel velvety, and extra feathery barbules help absorb the sound of their flight. The legs are covered in white feathers, and the longest central talon has small serrations that form a comb-like structure.

Mix of wing and body feathers SB

Mix of wing and body feathers SB

Leg and talons SB

Outer primary feather SB

Barn owl remains RM

Wings, wing and body feathers RM

Little owl
Athene noctua

Little owl wing feathers are relatively short, blackbird-length and broad with distinctive light and dark brown bars. As with the barn owl, the surface of the feathers looks and feels velvety. Upper body feathers are dark brown with large white or buff roundish markings. The legs are short and feathered, and often covered in dirt due to the owl's ground-feeding habits. The feathers on the legs are buff rather than the white of a barn owl, and the toes are sparsely feathered with stiff, hair-like structures.

Carcass

Leg sc

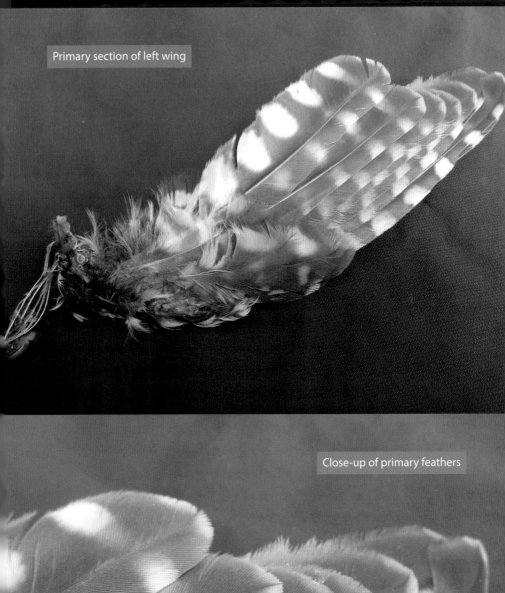

Primary section of left wing

Close-up of primary feathers

Swift

Apus apus

The primary feathers of swifts are very distinctive: long, narrow and with a black-green sheen on the outer webs and tips, and a smoky-brown colour on the inner webs. The inner primaries are four to five times smaller than the outer primaries. The secondaries are short and show the same colour pattern as the primaries. The inner webs bulge out beyond the tips of the outer webs to form a wavy edge. The tail feathers, particularly the outer ones, resemble shorter outer primaries.

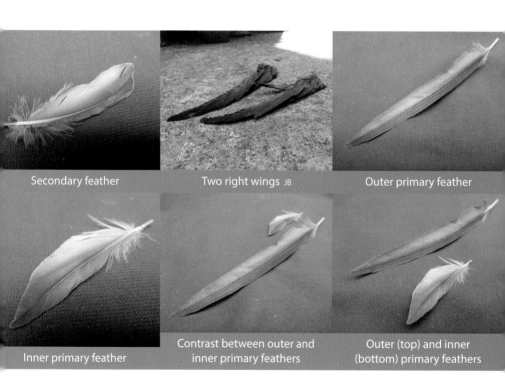

Secondary feather

Two right wings JB

Outer primary feather

Inner primary feather

Contrast between outer and inner primary feathers

Outer (top) and inner (bottom) primary feathers

Kingfisher
Alcedo atthis

RH

Kingfisher belly and breast feathers are generally bright orange, while the upper body plumage is made up of a range of spectacular blue feathers. The blue on many of the body feathers has a greenish tinge, while those on the back reflect an incredibly dazzling pale blue colour. The primaries and secondaries show pale blue on their outer webs. The white panels on the inner webs may have an orange tone. The upper and lower mandibles are often found: they are pointed, narrow and black with an orange-coloured sheath covering the inside. The lower mandible has a variable amount of buff/orange, particularly on females.

Primary feather

Breast feathers

Breast and scapular feathers

Lower (above) and upper (below) mandibles

Secondary section of
wing from above (top)
and below (bottom)

Lesser spotted woodpecker
Dryobates minor

VA

The wing feathers are tiny in comparison with great spotted woodpecker, being closer in size to of those of a large finch or bunting. They lack the red undertail feathers, and instead have white feathers with diagonal streaks. The back is striped and lacks the large white feathers that form white patches in the great spotted woodpecker. The shorter outer tail feathers are white and heavily barred. The feet are zygodactyl, which means that two toes point forward and two toes point back.

Half-eaten carcass at a goshawk nest, recorded by a nest camera GJ

Carcass showing barred undertail coverts

Great spotted woodpecker
Dendrocopos major

RH

Heads with a full red cap are juveniles; males have a small red patch at the back of the head. The silvery-grey upper mandible is often found detached and identified by its distinctive thick base which gradually tapers to a blunt tip. Red undertail feathers have a loose structure and fluffy appearance. The outer tail feathers are creamy-white with black spots or bars while the inner and central tail feathers are black. The tips are often worn and split. Feet are zygodactyl, with two toes pointing forward and two toes back.

Mix of primary feathers JB

Undertail covert

Head FH

Tail feather showing the split tip

Carcass in goshawk nest showing the zygodactyl feet JoM

Peregrine carrying a live great spotted woodpecker (juvenile) CS

Primary feather FH

Green woodpecker
Picus viridis

RM

Wing feathers are subtle in colour with bands of cream, green and khaki-brown. The outer webs of the secondaries and wing coverts are green. The inner webs have pale spots that disappear off the edge of the feather. The head is large and crow-size with red feathers on the crown. The yellow rump and dark striped flank feathers may also be found. As in other woodpeckers, the feet are zygodactyl, with two toes pointing forward and two toes back.

Two heads JB

Primary feather (underside) JW

Mix of wing feathers SB

Tail upperside (top) and underside (bottom)

Kestrel
Falco tinnunculus

The primary feathers are dark brown with contrasting regular white bars on the inner webs. In females and young birds these bars are tinted ochre-orange. The secondaries are dark brown with ochre-orange bars on their outer webs and lighter bars on the inner webs. In males the outer webs of some secondaries are black-brown and have reduced or absent colour bars. In males the tail feathers are blue-grey with black tips and dark bars towards their bases; in females they are orange-red-brown with dark bars and tips.

Primary feather, revealing the pattern on the inner web of a female or young bird

Secondary feathers and wing coverts of a female or young bird

Wing of a female or young bird

Primary feather (moulted) of a female

Ring-necked parakeet
Psittacula krameri

MSh

The bright apple-green feathers of ring-necked parakeets are unmistakable. Where this species occurs in feral populations it is a common feature in the diet of urban-dwelling peregrines. Any feathers of the parakeet will be very obvious and green. The under surfaces of the main wing and tail feathers appear yellow, while the upper surfaces are green. Their heads and brightly coloured large red beaks are commonly found. The cranium is often damaged and broken. The feet are zygodactyl, where two toes point forward and two toes point back.

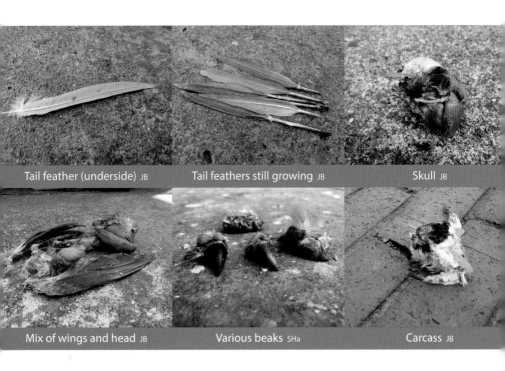

Tail feather (underside) JB   Tail feathers still growing JB   Skull JB

Mix of wings and head JB   Various beaks SHa   Carcass JB

Red-backed shrike
Lanius collurio

Red-backed shrikes have wing feathers that are redwing-size. The generally brown primaries have a thin line of red-brown running down their outer edge and a pale corner tip. The secondaries and tertials have a thicker band of red-brown running along the outer edge. The red-brown back/mantle feathers may also be found. Juvenile and female flank feathers are off-white with dark curved or V-shaped bars. These are flushed pink without any markings in males. Tail feathers are a striking black and white in males and red-brown in females and juveniles, with just a whitish or buff outer tail feather.

Back and flank (right) feathers (juvenile)

Primary feathers (juvenile)

Secondary feathers (juvenile)

Tertial feather (juvenile)

Great grey shrike

Lanius excubitor

VA

Great grey shrikes have striking black and white primaries and secondaries. The upper half of each feather is black and the lower half white. The tips of the secondaries are off-white. The head is distinctive with a grey cap, a thick black streak through the eye and ear coverts, and a hooked beak.

Wing DP

Head DP

Great grey shrike carrying jack snipe CH

Jack snipe placed on larder CH

Jay
Garrulus glandarius

RM

A jay's most obvious and striking feathers are the bright and dazzling blue and black striped wing coverts. The long, fluffy pink or white body feathers are often found in combination with tail and wing feathers. The primaries have a broad line of white on their outer webs and blue stripes towards their base. The secondaries are black with a large bright white patch along the outer web and some blue stripes towards the base. The tertials are a combination of black and a rich chestnut-brown. The tail feathers are black with blue stripes towards the base of the outer webs.

Head HS

Jay nest where adult has been predated leaving a pile of body feathers WB

Left wing and bones HS

Right wing JB

Right wing CS

Magpie
Pica pica

Magpies have very striking feathers. The primaries are black with an elongated white panel on the inner web. The secondaries are largely black with strong green-blue iridescence; a few of the outermost secondaries have some white on the inner web. The black tail is long and shows a range of iridescent colours, particularly blue-green, purple and hints of yellow. The white lower body feathers of the magpie are long and loose, with black bases.

Right wing cs

Primary feathers

Breast or belly feathers

Primary feathers

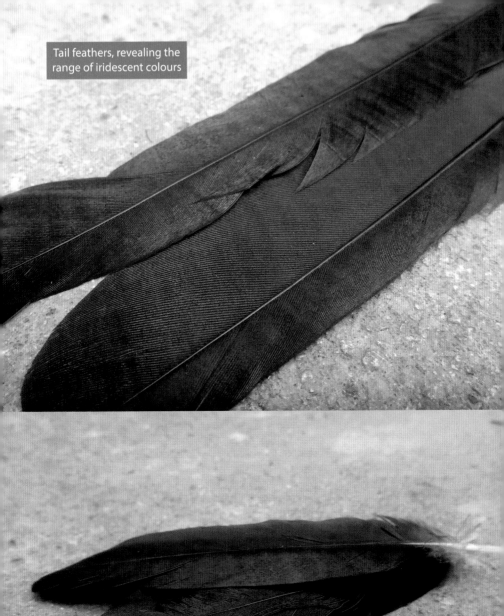

Tail feathers, revealing the range of iridescent colours

Inner secondary/tertial feathers

Jackdaw
Coloeus monedula

The remains of jackdaws are similar to those of other black corvids such as carrion crow *Corvus corone* and rook *C. frugilegus*. Young jackdaws are often targeted after fledging when they are vulnerable and their wing feathers are not fully grown. Although they have similarly shaped feathers to other corvids, their wing and tail feathers are shorter than in a crow or rook.

Wings JB   Primary feather   Leg

Tail feathers   Primary feather   Secondary feather

Head JB

Carcass JB

Blue tit

Cyanistes caeruleus

RH

A blue tit's primary and secondary feathers resemble those of the great tit, although they are smaller and have a more obvious blue-white edge to the outer web. The tertials and greater coverts have a broad whitish tip. The tail feathers are blue, and the outermost feathers may show some white on the outer web, though not bright white like those of the great tit.

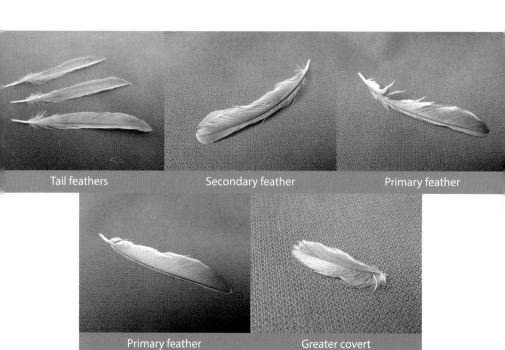

Tail feathers   Secondary feather   Primary feather

Primary feather   Greater covert

Great tit
Parus major

RH

A great tit's primary and secondary feathers have a subtle blue-white edge to their outer webs. The inner secondaries and tertials have broad cream-white tips. The outer webs of the dark tail feathers have a broader blue edge which becomes white in the outer tail feathers. The breast feathers are generally yellow with black bases.

Mix of feathers NR

Primary feathers

Tail feathers

Breast feathers

Inner secondary feather

Secondary feather

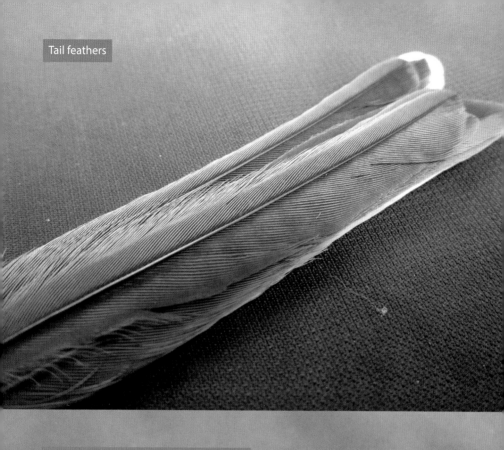

Tail feathers

Tail, revealing the white outer tail feather

Skylark
Alauda arvensis

Skylark feathers are starling-size. The primaries are predominantly dark brown, with light buff edges, and several of them have obvious emarginations. The inner primaries and the secondaries have elaborate light-coloured tips which form two bulges beyond the tip of the shaft. The tail feathers are similar to those of pipits or buntings. Juvenile skylarks show tertial feathers and wing coverts with obvious thick buff edges bordered on their inside with a dark outline.

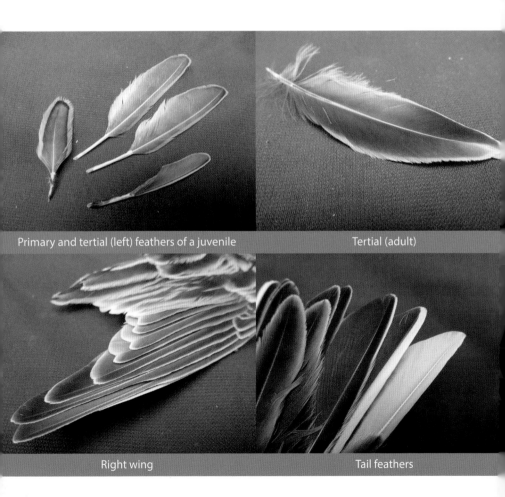

Primary and tertial (left) feathers of a juvenile

Tertial (adult)

Right wing

Tail feathers

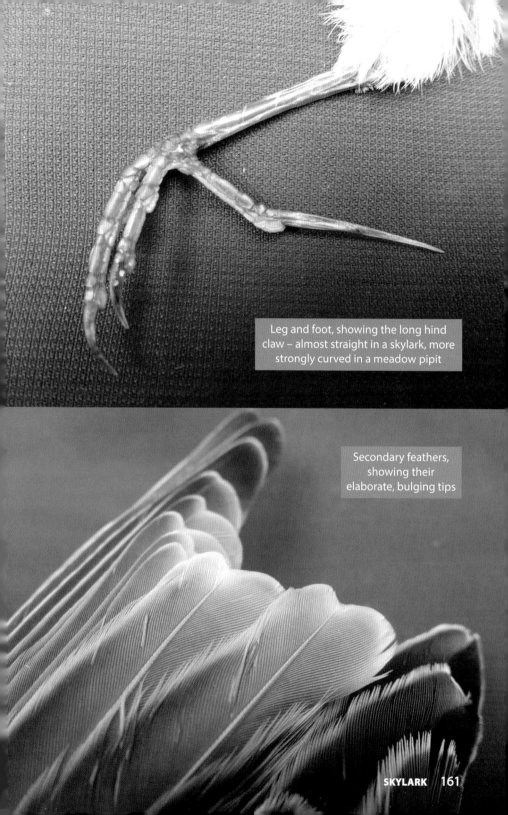

Leg and foot, showing the long hind claw – almost straight in a skylark, more strongly curved in a meadow pipit

Secondary feathers, showing their elaborate, bulging tips

Sand martin and house martin
Riparia riparia and *Delichon urbicum*

RH

House martins and sand martins have similar-sized wing and tail feathers. Their primaries are narrow and pointed, lacking the iridescent gloss found in longer swallow feathers. The sand martin has sandy-brown wing feathers. Those of the house martin are black-grey. If upper body feathers are found, they are sandy-brown in sand martin and black-grey with some purple-blue iridescence in the house martin.

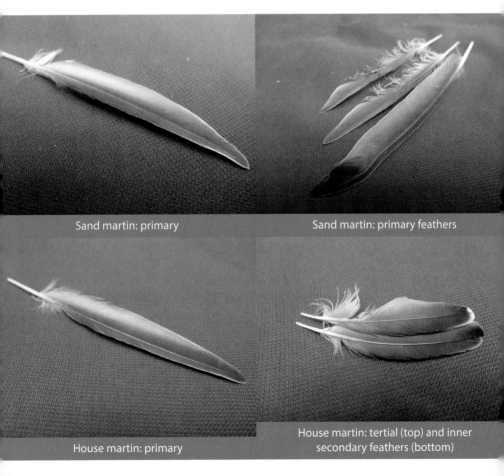

Sand martin: primary

Sand martin: primary feathers

House martin: primary

House martin: tertial (top) and inner secondary feathers (bottom)

House martin: primary section of wing

House martin: tail feather

Swallow

Hirundo rustica

RM

Swallows have long, narrow primary feathers with a green-black sheen, not unlike those of swifts. They are shorter than in a swift, however, with less contrast between the length of the outer and inner primaries. They also lack the smoky-brown inner webs of swifts. Instead the feathers appear blacker with black-grey inner webs. The breast feathers are buff-white. Tail feathers show single white spots or thick streaks on the inner web; outer tail feathers, particularly in adults, have long streamers.

Primary section of wing (juvenile)

Secondary feathers

Mix of wing feathers (juvenile)

Breast feathers

Outer tail feather (streamer)

Tail feathers

Primary feather

Iridescent wing coverts of right wing (adult)

Primary section of right wing (adult)

Willow warbler and chiffchaff

Phylloscopus trochilus and *P. collybita*

Both species have olive-green tail feathers and outer webs to their primaries and secondaries, with hints of yellow on the outer edge of the primaries. If a willow warbler or chiffchaff wing is found, the primary section can help identify it using the number of emarginations on the outer webs of the flight feathers. The willow warbler's three longest primaries (known as P3 to P5) each have an emargination, while in the chiffchaff it is the four longest primaries (P3 to P6) that are emarginated. If found plucked with an incomplete set of wing feathers, then either species may be involved.

Willow warbler RK   Chiffchaff HB

Chiffchaff: left wing   Chiffchaff: right wing   Chiffchaff: inner primary feathers

Chiffchaff: tail feathers

Chiffchaff: secondary feathers and greater covert

Blackcap
Sylvia atricapilla

RH

Blackcap primary feathers are dark grey and edged a lighter olive-grey; their inner webs have a thin white wedge running along their length. The longer primaries are emarginated and edged white. The outer webs of the secondaries, tertials and tail feathers are olive-grey with hints of brown.

Primary feathers

Tertial feathers

Secondary feather

Primary section of wing

Goldcrest
Regulus regulus

Goldcrest feathers are very small and easily missed. The outer edges of their primaries are tinged yellow and could be mistaken for chiffchaff or willow warbler. If inner primaries and secondaries are found, identification may be easier: the yellow edge on the outer web suddenly stops over halfway down, leaving the remaining outer web black-brown, to form a dark bar in the wing. The greater coverts are tipped buff-white and the tertials have obvious white tips.

Left wing

Secondary feathers and greater coverts

Primary feathers

Tail feather

Wren
Troglodytes troglodytes

RH

The feathers of wrens are very small and easy to miss. The outer webs of the primaries have buff-coloured bars against a darker background. The outer webs of the secondaries are a rich brown colour with dark alternating bars. The tail feathers are short, rich brown all over, and also have alternating dark bars.

Tail feather

Outer primary feather

Secondary feather

Outer primary feather

Starling

Sturnus vulgaris

While starling feathers come in lots of different patterns and colours, their red-orange legs, narrow tapering beak and pointed wing profile help with identification. Adults have dark primary feathers, thinly edged orange-buff with a pale buff shadow towards the tip, while young starlings have grey-brown feathers edged with a thicker orange-buff edge. Beaks are dark in non-breeding birds and develop a yellow sheath in late winter/early spring.

Head (winter) JB

Head (summer)

Wing feathers (juvenile)

Underwing (adult)

Wing (adult) DP

Primary section of wing (adult)

Body feathers (juvenile)

Primary (adult)

Skeleton of adult (spring/summer – yellow beak) cs

Legs, showing the strongly curved and relatively stout claws

Leg and half-eaten body (juvenile)

Secondary feathers and greater covert (adult) FH

Inner section of wing (juvenile) JB

Feather remains

Feather remains LW

Blackbird
Turdus merula

Male blackbirds have black wing and tail feathers with greyer inner webs. Females have fairly uniform dark brown wing and tail feathers, although the outer webs are usually a paler, warmer brown. Young birds are similar to females, but have orange flecks on their smaller wing coverts and scapulars. All have dark brown legs. An adult male generally has a yellow beak and yellow eye ring, though younger males are likely to have a fully or partially dark beak. A female and juvenile head will be dark brown with a dark beak, sometimes yellow-brown in adult females.

Wing and body feathers of predated female at nest WB

Tail feather (male)

Head (male) SC

Various feathers and leg (male)

Wing feathers (still emerging from sheath) and body feathers (juvenile) JB

Wing (male) SC

Carcass (male) CS

Wing and body feathers (juvenile)

Fieldfare

Turdus pilaris

Fieldfares have easily identifiable breast and flank feathers. These have distinctive black curved bands or V-shaped markings against a white background. Those from the breast tend to have burnt-orange-brown tips. Their legs are black, and the tail feathers are black with grey tones along the outer webs. The primary feathers are like those of blackbird but longer and paler, often with a faint pale or white line along the outer edge. The outer webs of the secondaries are blue-grey with bright rich brown edges.

Primary section of wing

Tail feathers

Lower body feather

Undertail covert

Secondary feather

Greater coverts

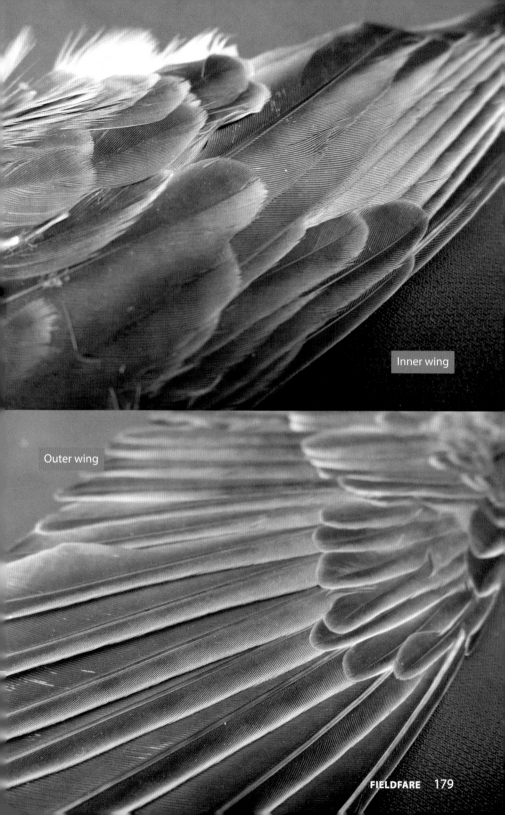

Inner wing

Outer wing

Redwing
Turdus iliacus

GT

While similar to the song thrush, redwings are distinguished by a number of distinctive features. The head has a creamy-white eye-stripe and the lower mandible is yellow (pale off-white in the song thrush). The wing feathers are greyer-brown, often olive, although in some individuals will appear brighter. The patches towards the bases of the inner webs are washed out and pale or cream, lacking the brighter orange-yellow of the song thrush. The flank feathers are distinctively orange-red. The breast is whiter than in song thrush, and appears streaked rather than spotted.

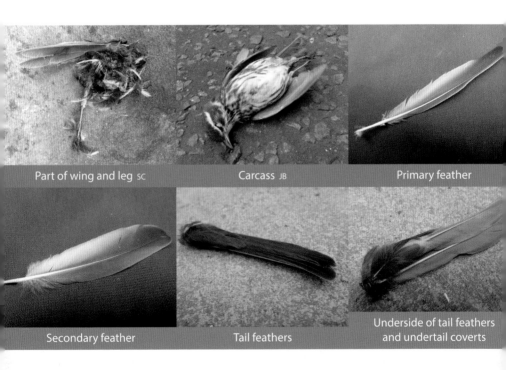

Part of wing and leg SC

Carcass JB

Primary feather

Secondary feather

Tail feathers

Underside of tail feathers and undertail coverts

Song thrush
Turdus philomelos

RH

Song thrushes have bright, rich brown wing and tail feathers which during the winter months can easily be confused with those of redwing. In the song thrush the lighter colour patches on the inside web of the primaries and secondaries are a brighter orange-yellow colour. Their greater coverts have similarly coloured tips, often rather indistinct in an adult but very clear, and shaped like rose-thorns, in a juvenile. The flanks lack the red-orange feathers of redwing. The spots on the breast and belly are bolder and better defined; in the redwing they appear as streaks or thick lines.

Body and wing feathers SG

Breast feathers

Wing (underside)

Secondary feathers and greater coverts, showing 'rose-thorn' tips of juvenile plumage

Carcass of song thrush DP

Carcass of redwing AL

Mistle thrush
Turdus viscivorus

The mistle thrush's wing feathers do not obviously look like those of a thrush and can mislead identification. The primaries and secondaries are longer than those of blackbird or fieldfare, sandier in colour and with white or creamy patches towards the bases of the inner webs. The tail feathers are sandy in colour and have white tips, with the outermost feathers having white along the length of the outer web. Other thrush species lack these white markings. The belly and breast feathers have obvious bold round black spots and are often found in small clusters on the ground.

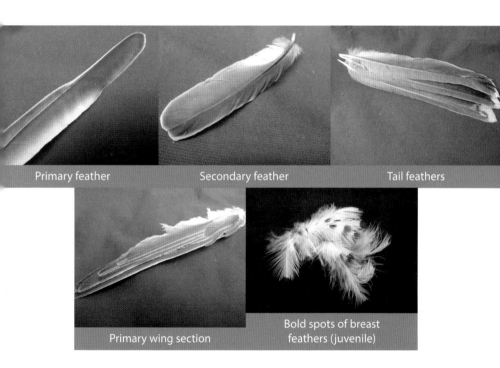

Primary feather   Secondary feather   Tail feathers

Primary wing section   Bold spots of breast feathers (juvenile)

Robin

Erithacus rubecula

Robins have a light olive ochre-brown tinge to the outer webs of their primary, secondary and tail feathers. The pale panels along the outer edges of the inner webs have washed-out hints of ochre-brown rather than being white or pale grey as in many other small passerines. If sections of the wing are found, orange-cream tips may be present on the greater coverts; 'rose-thorn' tips indicate a juvenile or first-year bird, while in adults the lighter tips of the greater coverts are indistinct or often absent. Their orange-red breast feathers may also be found.

Inner primary feather

Tail feather

Wing showing greater coverts, with 'rose-thorn' tips of juvenile plumage

Inner primary feather

Wheatear
Oenanthe oenanthe

RM

The feathers of a wheatear are similar in size to those of skylark and starling, and are easily mistaken for those species at first glance. The presence of black and white tail feathers helps to confirm the identity. In these, the upper third to a half is black with buff edges, and the lower portion is white. The primaries are sandy-brown with a thin buff edge; the tips are darker brown with a thicker buff edge. The secondaries are also sandy-brown with a thicker buff outer edge. The black legs, if found, are also a useful clue.

Secondary feathers

Inner primary feather

Head HS

Middle primary feather

Selection of wing and tail feathers

Close-up of tail feather

House sparrow
Passer domesticus

House sparrow wing feathers have a distinctive buff edge to their outer webs and a sandy-buff patch towards the base of the outer webs. On the inner primaries and secondaries, the buff edge is more extensive, leading to the tip. The tail feathers are dark sandy-brown with thin buff outer edges. Male house sparrows are more varied in colour with chestnut-brown greater coverts, sandy-brown to chestnut-brown edges to their tertials and secondaries, and white median coverts.

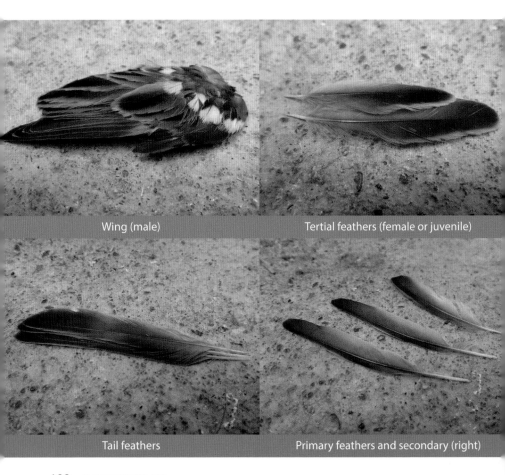

Wing (male)

Tertial feathers (female or juvenile)

Tail feathers

Primary feathers and secondary (right)

Dunnock
Prunella modularis

Dunnock primary and secondary feathers resemble those of house sparrow and reed bunting at first glance. The reed bunting has a thicker, brighter white patch on the inner web and a more defined, deeper brown outer web. The edge of the dunnock's outer web is orange-brown, not buff as in the house sparrow, and does not broaden towards the base of the feather. Greater coverts are dark brown with a thick band of orange-brown on their outer webs and small cream tips separated by the darker central shaft.

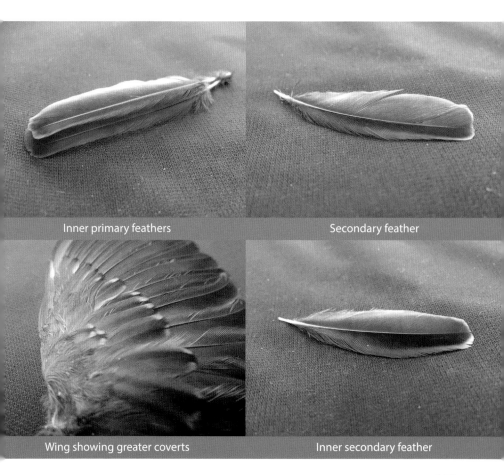

Inner primary feathers

Secondary feather

Wing showing greater coverts

Inner secondary feather

Pied wagtail
Motacilla alba

Pied wagtails have comparatively plain and uniform primaries and secondaries, despite their characteristic black and white tail and tertial feathers. Their primaries and secondaries are dark grey rather than brown, with a white wedge on the lower portion of the outer edge of the inner web. A faint white or light grey edge catches just the edge of the outer web. The long, narrow tail feathers, if present, help confirm identification. The central tail feathers are black and often with a faint white or light grey edge. The outer tail feathers are mainly white, with some black towards the base and on the inner web.

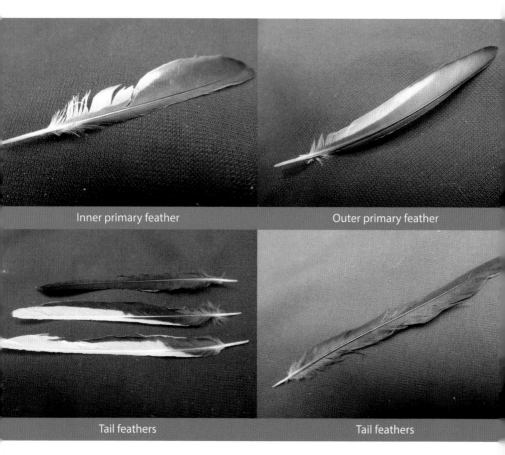

Inner primary feather

Outer primary feather

Tail feathers

Tail feathers

Tail feathers

Tail (left) and wing feathers

Meadow pipit
Anthus pratensis

RM

Both the meadow pipit and the tree pipit have relatively uniform, plain-coloured wing feathers, with greater coverts that are tipped creamy-white. Their primaries and secondaries are sandy-brown with a yellow-cream edge, and more yellow if fresh. Their tail feathers are a similar colour apart from the outermost feathers, which are mainly white with brown on the lower portion of the inner web. If legs and toes are found, a meadow pipit has a very long hind claw: 10–13 mm, compared with 7–9 mm in tree pipit.

Secondary section of wing

Outer tail feather

Leg and foot, showing the long hind claw – longer and straighter than tree pipit, but more curved than skylark

Breast feathers

Tree pipit
Anthus trivialis

For description, see meadow pipit. If the whole wing of a tree pipit is found the fifth primary is shorter than that of a meadow pipit: 2–6.5 mm shorter than the wing tip, compared with 0–1 mm, sometimes 2 mm, shorter in meadow pipit.

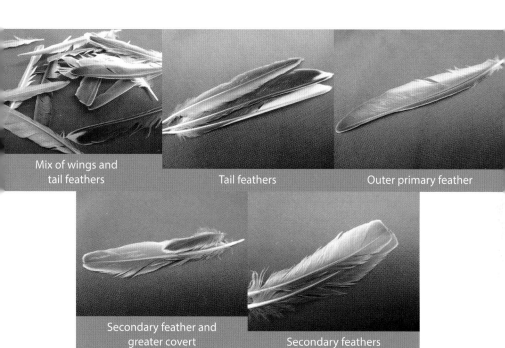

Mix of wings and tail feathers

Tail feathers

Outer primary feather

Secondary feather and greater covert

Secondary feathers

Chaffinch
Fringilla coelebs

RH

Chaffinches have distinctive primary and secondary feathers with partial yellow edges on the upper portion of their outer webs and white patches towards their bases. These markings are reduced or absent on the outer primaries. The greater coverts are tipped creamy-white. The tail feathers each have a different spread of white markings, with the outer ones being most white. The central tail feathers are dark, with grey-blue-green hues. The male's dull pink breast feathers and both sexes' green rump feathers may also be found, particularly if recently plucked.

Tail feathers

Secondary section of wing

Secondary (top) and primary (below)

Secondary feathers and greater covert

Primary feather

Pink breast feathers (male)

Green rump feathers (male)

Tertial feather

Secondary feather JB

Mixed wing and tail feathers (sparrowhawk kill) SG

Brambling
Fringilla montifringilla

At first glance a brambling's primary and secondary feathers resemble those of a chaffinch. However, the edges of their outer webs have hints of orange rather than white or yellow. The broad white patches at the base of their outer webs are bordered with black beneath. The greater coverts are tipped buff-orange. The tail feathers contain wedges of a smoky-grey colour rather than the larger patches of white in the chaffinch.

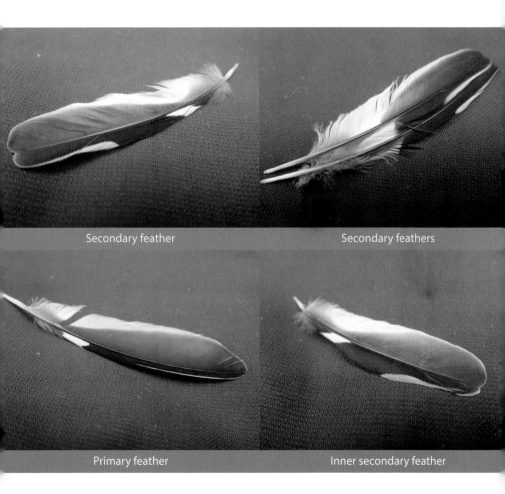

Secondary feather

Secondary feathers

Primary feather

Inner secondary feather

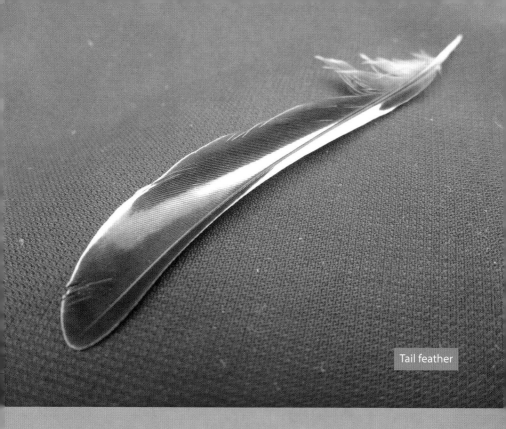

Tail feather

Greater covert

Hawfinch
Coccothraustes coccothraustes

RH

Everything about a hawfinch is striking, from its huge beak to its black and white wing feathers. The primaries have extravagant bulging and angled iridescent tips, while the secondaries broaden towards the tip. Their inner webs have bright white portions. The beak is thick and large; it is often snipped off by a raptor and discarded.

Primary feather

Inner primary feather

Secondary feather

Beak PJ

Bullfinch
Pyrrhula pyrrhula

The primary feathers of a bullfinch look uniform and grey at first glance. A closer look reveals a subtle darker outer web which is iridescent – and this is more obvious and extensive on the secondaries and tertials. The innermost tertial feather is iridescent on the outer web, and the inner web is bright pink on a male or pink-grey on a female. The blue-purple iridescence is extensive across the outer webs of the black tail feathers, and all across the central tail feathers.

Small wing feathers including the pink inner tertial (male)

Mixed wing feathers, undertail coverts and pink inner tertial (female)

Secondary feather

Primary feathers

Greenfinch
Chloris chloris

RM

Greenfinch feathers are longer than those of goldfinch. The black-grey of the primaries does not contrast as strongly with the pale yellow outer webs. The secondaries lack the yellow bars of goldfinch and are fringed yellow-green with grey tips. Male primaries and secondaries are bright yellow and grey, the yellow extending right across the outer web of the outer primaries. Females have less extensive yellow, which does not reach the shaft. Their tail feathers have variable amounts of yellow with grey-black portions towards their tips. The central tail feathers are a blend of grey and green-yellow all over. Their thick, strong beaks are often removed and found amongst prey remains.

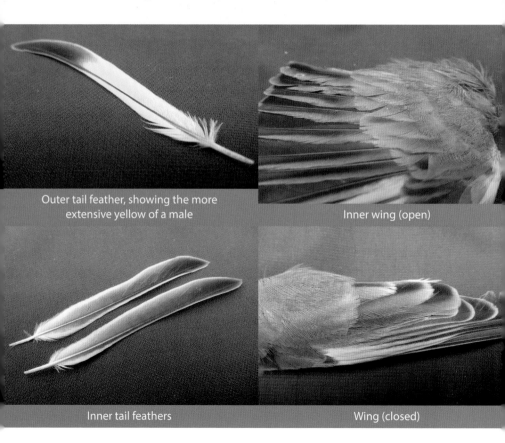

Outer tail feather, showing the more extensive yellow of a male

Inner wing (open)

Inner tail feathers

Wing (closed)

Body feather

Secondary feather

Middle primary feather

Skull

Outer primary feather (male)

Primary section of wing

Linnet
Linaria cannabina

Linnets have primary and tail feathers with thick white outer edges, which almost cover the whole outer web in some tail feathers. The outer webs of the secondaries are grey-brown to buff with a white-edged tip. The tips of the primaries are edged buff-white. The male's red breast feathers may also be found amongst buff-white body feathers and rich brown back feathers.

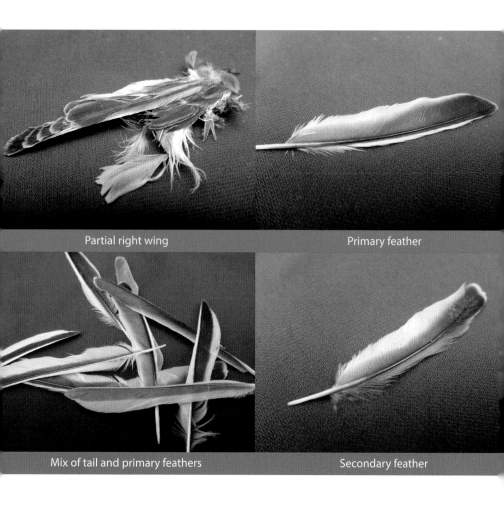

Partial right wing

Primary feather

Mix of tail and primary feathers

Secondary feather

Goldfinch
Carduelis carduelis

Goldfinch feathers stand out amongst prey remains and the bright yellow on their primaries and secondaries is striking and unmistakable. The yellow outer webs contrast with the black of the rest of the feather, and is accompanied by a white-edged tip. The tail feathers are black with white tips, although often the latter are worn away. The similar greenfinch feathers are longer and less contrasting.

Primary feathers

Primary feather

Tail feathers

Inner secondary feather

Tail feathers (juvenile)

Secondary feather

Siskin

Spinus spinus

Siskin primary and secondary feathers are goldfinch-size. The primaries are grey-black with a thin yellow line along the outer web and a thicker yellow patch or band towards the base. The pale portion of the inner web is a weak lemon-yellow colour. The secondaries have a larger band of yellow towards the base of their outer webs and a thicker yellow edge towards their tips. Siskin tail feathers are generally two-thirds yellow with a darker black-grey outer section that tapers down towards the shaft. The central tail feathers are dark all over and edged yellow.

Outer primary feather

Tail feathers

Tail feather

Left wing

Yellowhammer
Emberiza citronella

The yellowhammer's primary, secondary and tail feathers may resemble other species such as chaffinch, meadow pipit and house sparrow (depending on which feather). Combined clues such as the yellow edges to the primary and tail feathers, yellow and rusty-brown body feathers and extensive orange-buff wedges in the tertials may help confirm identification.

Inner wing

Outer wing

Tail feathers

Rump and tail feathers

Reed bunting
Emberiza schoeniclus

Compared to house sparrows, reed bunting wing feathers are richer in colour, with thicker bands of chestnut-brown on their outer edges. The primaries lack the sandy-brown patch towards their base and instead have a continuous chestnut edge leading into white towards the tip. The secondaries have a thick chestnut-brown edge, which is darker and deeper chestnut than in a male house sparrow. The median coverts lack the white or buff of house sparrows and the greater coverts are tipped white. The central tail feathers have a chestnut-brown outer web and a black inner web. The remaining tail feathers are either black or black with portions of white running along their inner and outer webs.

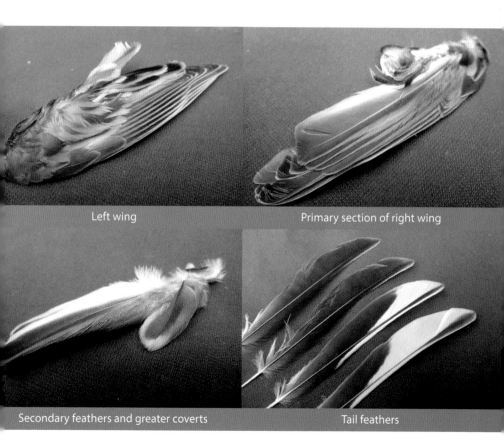

Left wing

Primary section of right wing

Secondary feathers and greater coverts

Tail feathers

Fledglings

The feathers from fledglings of small birds such as great tits and robins are often found along woodland footpaths and in parks and gardens after they have been predated by raptors, especially sparrowhawks. Their wing, tail and body feathers are often found together in small piles in the same area. The feathers may not be fully developed, showing the blue-grey sheaths or pins from which they emerge.

Great tit

Great tit

Blackbird

Dunnock

Other animals

Small animals such as voles, mice, bats, shrews and amphibians, such as toads, provide important food for many raptors. Their carcasses are often cached in nest boxes, while their remains may be found in nests, beneath roost or nest sites and in pellets. Larger mammals, such as rabbits, hares, grey and red squirrels, weasels and stoats, are easier to identity using signs of fur, whole or partial carcasses and skulls. This section shows some of the typical carcasses and remains that raptors leave behind or have at their nests.

Toad JB

Bat (species unknown)

Grey squirrel tail

Frogspawn

Tip of grey squirrel tail, showing the typical orange hair MS

Remains of grey squirrels in a goshawk nest MS

Remains of grey squirrels in a goshawk nest SE

Fresh skull of grey squirrel in goshawk nest TB

Fresh lower jaw of grey squirrel in goshawk nest TB

Fresh, partly plucked grey squirrel in goshawk nest TB

Old grey squirrel skull beneath a goshawk nest

Fresh rabbit skull and fur

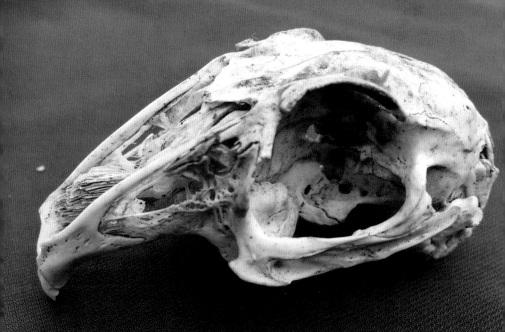

Rabbit skull

Brown hare skull

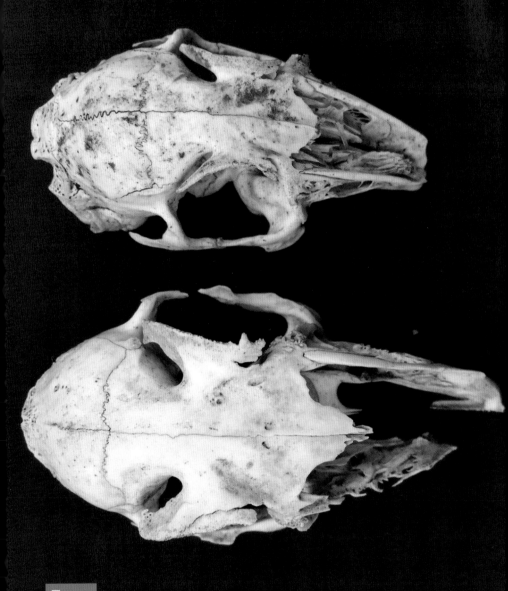

Brown hare skull (left) and rabbit skull (right)

Buzzard chicks with cached rabbits AF

Rabbit

Rabbit and red-legged partridge remains in a goshawk nest MS

Goshawk chick on nest with half-eaten carrion crow MS

Rabbit fur and skull

Peregrine chicks with cached noctule bats (Hungary) MP

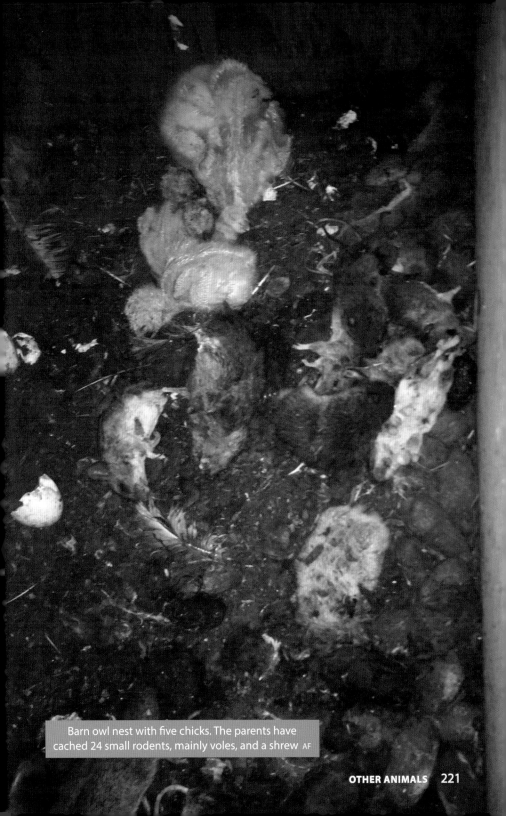

Barn owl nest with five chicks. The parents have cached 24 small rodents, mainly voles, and a shrew AF

Red kite *Milvus milvus* taking a fish from water RT

Buzzard carrying an adder RT

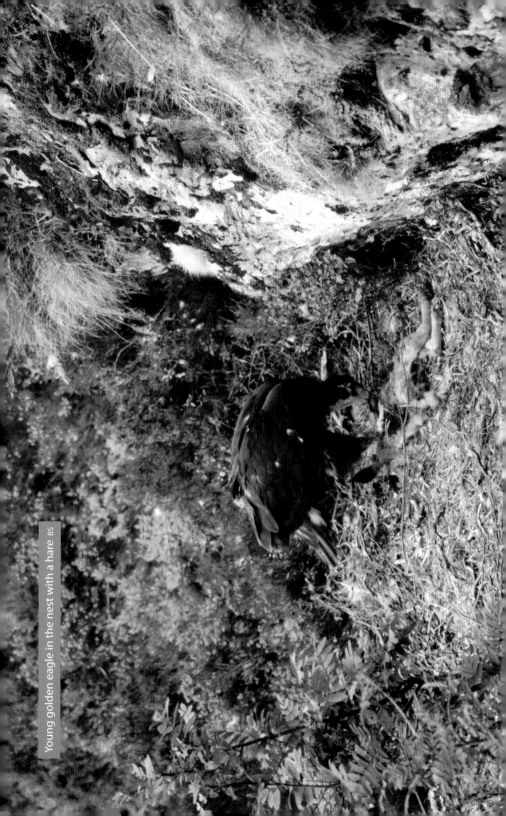

Young golden eagle in the nest with a hare BS

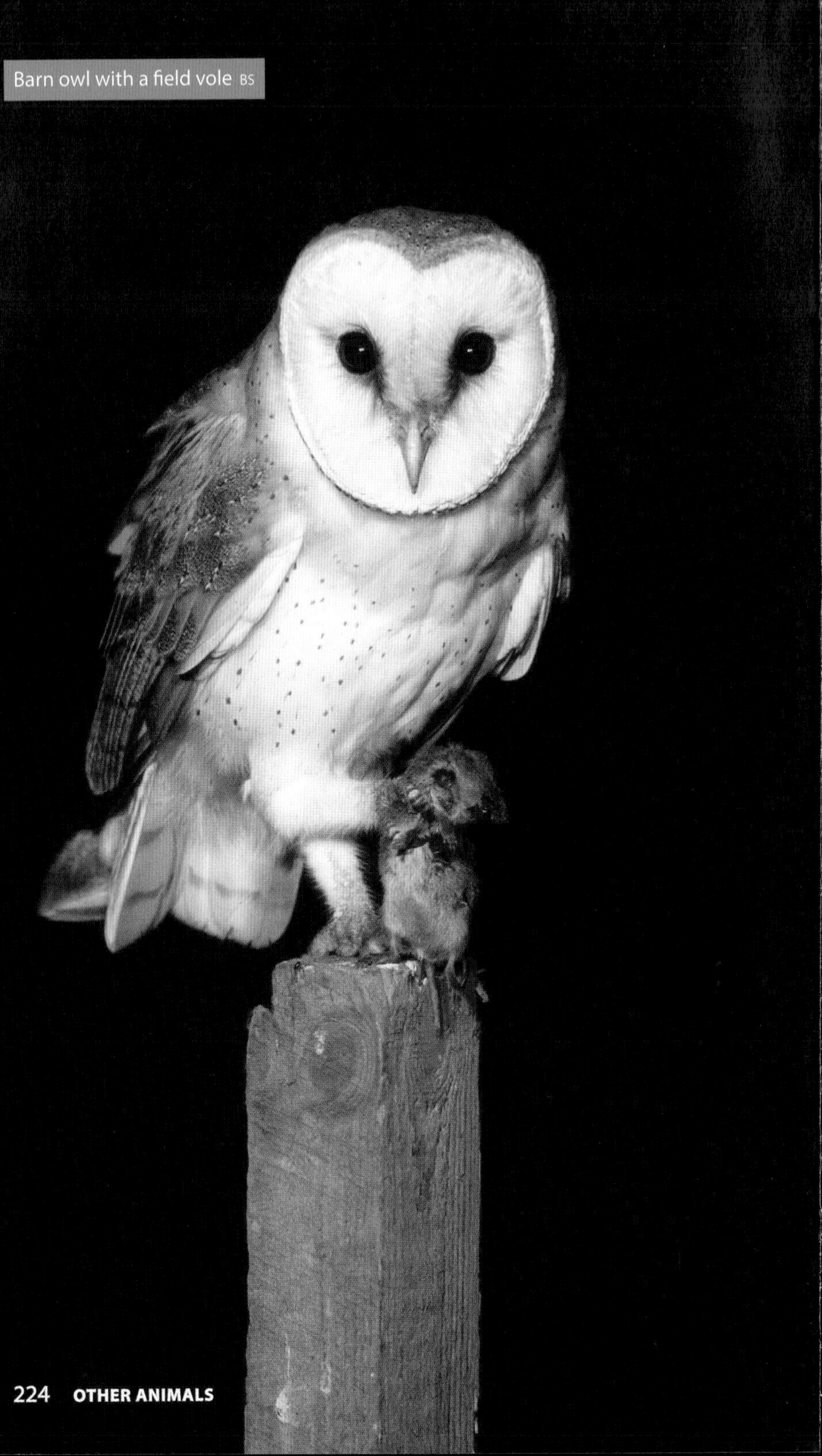

Barn owl with a field vole BS

Juvenile peregrine carrying feral pigeon prey JWa

SCIENTIFIC NAMES OF OTHER ANIMALS MENTIONED IN THE TEXT

Adder	*Vipera berus*
Frog	*Rana temporaria*
Toad	*Bufo bufo*
Noctule bat	*Nyctalus noctula*
Red fox	*Vulpes vulpes*
Badger	*Meles meles*
Stoat	*Mustela erminea*
Weasel	*Mustela nivalis*
Grey squirrel	*Sciurus carolinensis*
Red squirrel	*Sciurus vulgaris*
Field vole	*Microtus agrestis*
Mountain hare	*Lepus timidus*
Brown hare	*Lepus europeaus*
Rabbit	*Oryctolagus cuniculus*

Photo credits

Photos without a credit beside them have been taken by the author. Collating such a huge collection of photographs is not an easy task and would not have been possible without the amazing contributions made by all the photographers listed below.

AF	Anna Field
AL	Adrian Ludlam
AR	Alistair Ray
ARo	Adam Rogers
BL	Brian Lancastle
BS	Bobby Smith
CH	Chris Hurst
CL	Colin Lea
CR	Craig Reed
CS	Chris Skipper
DP	Dave Pearce
FH	Francis Hickenbottom
FV	Francesco Veronesi (Flickr, CC BY-SA 2.0)
FVa	Frank Vassen (Flickr, CC BY 2.0)
GC	G. Craggs
GG	George Gay
GJ	Gareth Jones
GT	Gary Thoburn (garytsphotos.zenfolio.com)
HB	Flickr/hederabaltica (Flickr, CC BY-SA 2.0)
HOT	Hawk and Owl Trust
HS	Hamish Smith
JB	John Boorman
JC	James Chubb
JD	Jack Devlin
JH	John Hansford
JHu	Jackie Huxtable
JM	Jane Memmott
JMa	H.S. John Madge
JoM	Josh Mitchell
JT	Jason Thompson (Flickr, CC BY 2.0)
JWa	Jon Watson
JW	Janet Wickings

LS	Luke Sutton
LW	Lizzie Wilberforce
MD	Matt Davis (Flickr, CC BY 2.0)
MJ	Mike A. Jackson
MK	Mike King
ML	Michael Leigh-Mallory
MP	Mátyás Prommer
MPi	Matthew Pitcher
MS	Mark Sharples
MSh	Mary Shattock (Flickr, CC BY-SA 2.0)
NG	Nick Gates
NR	Nat Roberts
OK	Owen Kirby
PB	Paul Bowerman
PJ	Phil Jones
PR	Pete Richman (Flickr, CC BY 2.0)
RB	Ryan Burrell
RH	Rod Holbrook
RK	Ron Knight (Flickr, CC BY 2.0)
RM	Robin Morrison
RT	Richard Tyler (richardtyler.zenfolio.com)
RV	Rupal Vaidya (Flickr/Vaidyarupal, CC BY 2.0)
SB	Sophie Bagshaw
SC	Saimon Clark
SE	Simon Evans
SG	Stephen Gilliard
SH	Sam Hobson (samhobson.co.uk)
SHa	Stuart Harrington (London Peregrine Partnership)
TB	Tim Bray
TV	Terence Voller (Flickr, CC BY 2.0)
VA	Vitalie Ajder
VM	Venetia Manning
WB	William Bick
WL	Will Langdon

With thanks

I am very grateful to everyone who has contributed images for this book. Bringing together such a huge range of raptor prey remains to create this visual resource has been no mean task, and has only been possible because of the generosity of so many people who devote their time to watching, observing and being passionate about wildlife – and raptors in particular.

A huge thank-you also to my wife Liz, who has supported me along the way, encouraged me throughout, and put up with innumerable unsavoury items scattered around the house and garden.